THE
ULTIMATE FACT BOOK
. . . .

THE ULTIMATE FACT BOOK

COUNTRIES A TO Z

Andrew Wojtanik

Developmental Studies Center

To my sister, Jessica,
who sees the world in a special way

Acknowledgments:
There are a number of people I would like to thank who have been supportive for so long and have given me the motivation to reach my goals:
- my parents, Daniel and Dianna Wojtanik, who have been behind me every step of the way ever since the day I was born and who have given valuable advice throughout my life
- my grandparents, Raymond and Ruth Rippley, who also backed me one hundred percent and encouraged me to pursue my goals and accomplish them
- my entire family living in the Kansas City area, who have been extremely supportive, especially Denise Brame, who has given me many important geography books
- my aunt and uncle, Virginia and Edward Pryze, who gave me my first subscription to *National Geographic Kids* magazine
- all of my relatives living in western New York, who cheered me on as I competed in the National Geographic Bee. To this day, I am a die-hard Buffalo Bills fan because of them (yes, this *is* a good thing!)
- Mr. and Mrs. Vince Stonestreet, who provided the best world atlases I have ever owned
- the Lakewood Middle School community, including Mr. Scott Currier, who went out of his way to support me while I was competing in the Geographic Bee
- Mrs. Julie Larsen, my sixth-grade social studies teacher, who shared with me her tremendous love for geography
- my friends who rooted for me in the Bee and challenged me to compete at the highest level.

Lastly, I need to acknowledge National Geographic for giving me the opportunity to share my knowledge of the world. Special thanks goes to David B. Miller, Geoffrey W. Hatchard (the runner-up in the 1992 National Geographic Bee who now works for NGMaps!), Verónica E. Betancourt, David M. Seager, Andrew Murphy, Suzanne Fonda, and the rest of the National Geographic staff for their hard work and dedication in making this project a reality.

Text copyright © 2005, 2012 Andrew Wojtanik
Maps copyright © 2005, 2012 National Geographic Society
Compilation copyright © 2005, 2012 National Geographic Society
All rights reserved.

This Developmental Studies Center edition is designated for the educational market only and was published by arrangement with the National Geographic Society.

The text type is set in New Caledonia; headlines are set in Clarendon

Developmental Studies Center
1250 53rd Street, Suite 3
Emeryville, CA 94608-2965
800.666.7270 ○ fax: 510.464.3670
devstu.org

ISBN 978-1-61003-380-0
Printed in Hong Kong

2 3 4 5 6 7 8 9 10 RRD 20 19 18 17 16 15 14

Note from the Author	7
Things You Need to Know	9
World and Continent Maps	11
COUNTRIES A–Z	19
Geographic Extremes	378
Glossary	380
Bibliography	382
About the Author	383

Note from the Author

I have watched seven National Geographic Bees since I won the contest in 2004 and published the first version of this book. Yet while watching the Bee each May, I cannot resist feeling a bit of nostalgia. In four years of high school and three years at Georgetown University in Washington, D.C., my experiences with the National Geographic Bee—a geography competition that boasts more than five million fourth- through eighth-grade participants across the country every year—have remained an integral part of my studies and my ambitions.

It began with an atlas . . . and a hint of curiosity. What was it about geography that originally captivated me? Maybe it was my tendency to get lost in visions of far-away places while examining maps of different regions of the globe. Perhaps it was the satisfaction I gained from planning family vacations. Or possibly, it was the cultures and histories of these interesting places, far from my native Kansas, that captured my imagination. Most likely, it was a combination of all of these factors that sparked my interest in the study of the world's features.

Along came the National Geographic Bee when I was in the sixth grade. After I won my first Bee "title" at the school level, my fixation with geography intensified. I put up a good fight at the state competition in Hays, Kansas, but fell short of moving on to the next level.

My growing knowledge and confidence, however, produced a better result the following year. I advanced to the national competition in Washington, D.C., where I met other students who enjoy geography as much as I do. Alas, I lost in the preliminaries and watched the finals from the audience rather than as a contestant on stage.

It wasn't until my third and final try (I was an eighth grader) that I achieved my goal of reaching the "top ten" at the national competition. Just proceeding that far was an extraordinary achievement for me, but the excitement did not quite end there. After answering questions about Walvis Bay (Namibia), La Amistad National Park (Costa Rica/Panama), the Heard & McDonald Islands (Australia), and many others, I was declared the champion of the 2004 National Geographic Bee. It was an accomplishment that exceeded all expectations but could not have been achieved without incredible dedication and work ethic.

When Alex Trebek, host of the hit game show *JEOPARDY!* and moderator of the Bee, asked me how I studied for the contest, I noted that I had created a 400-page study guide. This monstrous packet of research consisted of information about the physical, political, and environmental features for every country in the world (at the time, 192 of them). This guide, along with a decent memory and a little luck, were major contributors to my success.

Three months later, when I thought my string of surprises had ended, National Geographic offered to publish my geography fact book. The result was the edition preceding the book you now hold in your hands.

It is with great privilege that I am able to share the updated edition of this book as well as my love for geography with you. I encourage you to get involved in the Geographic Bee if you are the right age, and I hope this fact book will be useful on your fantastic journey. I also recommend obtaining a detailed atlas for visual reference so you can find the exact location of the features listed in this book.

When competing in the Geographic Bee, however, keep in mind that you don't have to advance to the national finals to achieve the satisfaction of gaining a wider knowledge of the world. With a willingness to expand your scope of study to places you may have never seen, I hope that you will find geography—spanning and incorporating political, economic, environmental, and cultural fields—as eye-opening as I do. After all, the Geographic Bee comes and goes, but the sometimes inexplicable desire to delve into the intricacies of foreign lands will forever live on.

So, whether you are preparing for the National Geographic Bee or simply would like to add to your geographical repertoire, *The Ultimate Fact Book* is the reference guide for you!

Best of luck!
Andrew Wojtanik

Things You Need to Know

Geography is a huge subject, as you probably have already discovered for yourself. It would be impossible to list every fact about every country. So I had to make some choices when I put my book together. I selected key things that I thought would help me study for the National Geographic Bee. Then I organized them in a way that made sense to me.

First, I listed each of the world's 195 independent countries in alphabetical order. Then I decided what facts I needed to know about each and organized them into three categories: physical, political, and environmental/economic. Within each of these categories I listed topics that fit the category and put these in the same order for each country. Of course, if a topic didn't apply to a country, I didn't list it. For example, you won't find rain forests listed for Iceland.

I was selective in what I listed for each country. In the case of physical features and political regions, only major ones are included. For some countries a mountain range will be named but no peaks will be singled out. That's because I only listed peaks that are at least 8,000 feet (2,400 meters) high. In countries that have several peaks that exceed this elevation, I listed the ones that I thought are the most significant or well known. The same is true for rivers. In some cases you will see rivers on the country map but no rivers listed in the text. That's because these rivers are usually less than a hundred miles long. For depressions, only those that are at least 50 feet (15 meters) below sea level are listed.

For the most part, features and places are listed in alphabetical order. The major exception is cities. These are listed in order of population, from largest to smallest. I only listed capital cities, a country's largest city, or cities that had at least a million people.° Ethnic/racial groups are listed largest to smallest; religions are listed by number of followers, largest to smallest; and languages are listed in order of the number of people who speak them, largest to smallest. The top natural resources and major exports are listed in order of importance, as are the leading agricultural products.

There are two "Region" categories, one in physical and one in political, so don't think this is an error. Also, the heading "Independence" does not necessarily refer to when a country became free from control by another country. It generally marks the commonly accepted date for when the country became an independent political entity. Finally, most accents and diacritical marks have been intentionally omitted from the text, although they do appear on the maps.

All of the maps in this book were created by the cartographers at National Geographic as part of the Xpeditions Atlas, an online collection of maps intended for use by students. (The maps can be printed or downloaded from the Society's Web site: www.nationalgeographic.com/xpeditions/atlas/. The same Web site also offers blank outline maps of each country.) In some cases, places that I mention in my text are not labeled on the country map. Likewise, some places on the country map are not included in my text, and sometimes the name on a map is different from what I have in my text. That's because I use the commonly accepted international place-names, while the map may use the local spelling. Remember, to get the maximum benefit from this study guide, you will need to supplement it with up-to-date atlases and other reference books.

°In the 2012 update of this book, for countries with numerous large cities, we only listed cities that met a higher population threshold (e.g., two to three million people).

THE WORLD

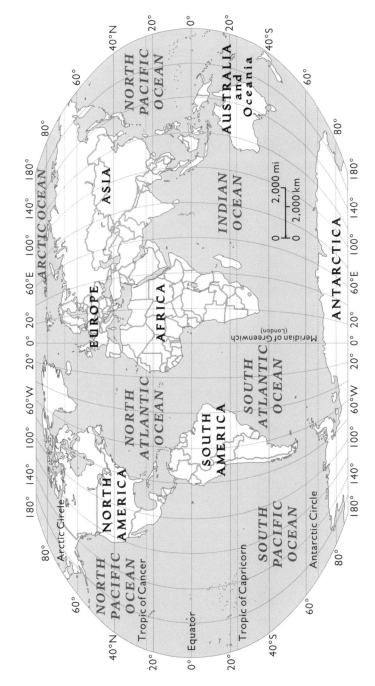

WORLD AND CONTINENT MAPS | PAGE 11

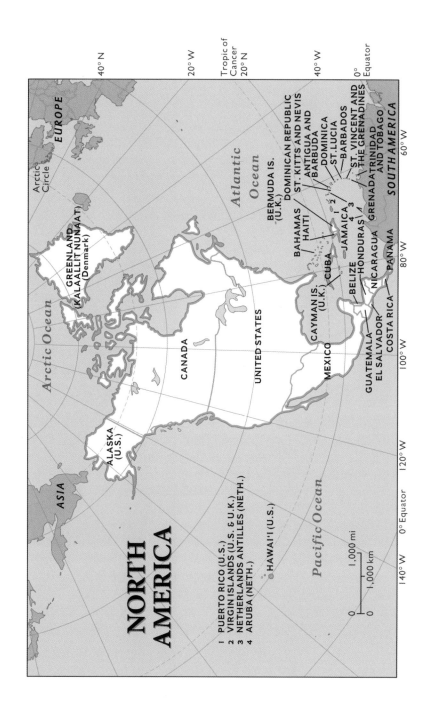

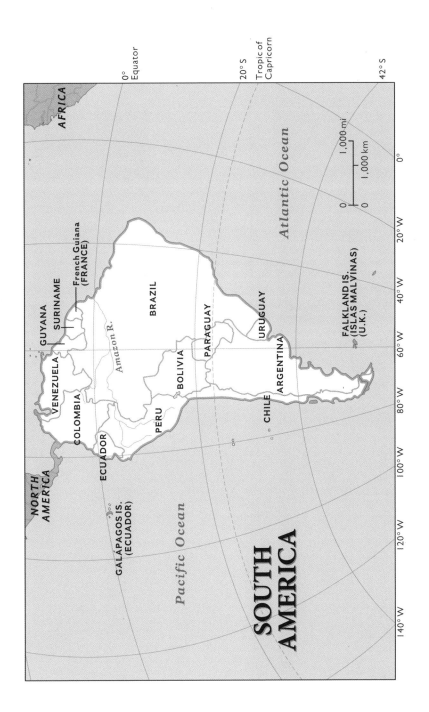

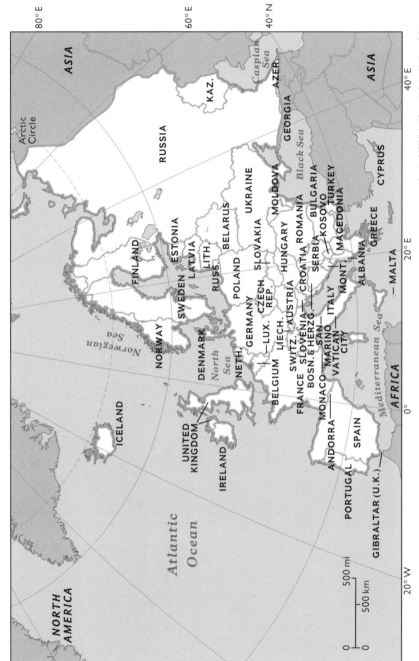

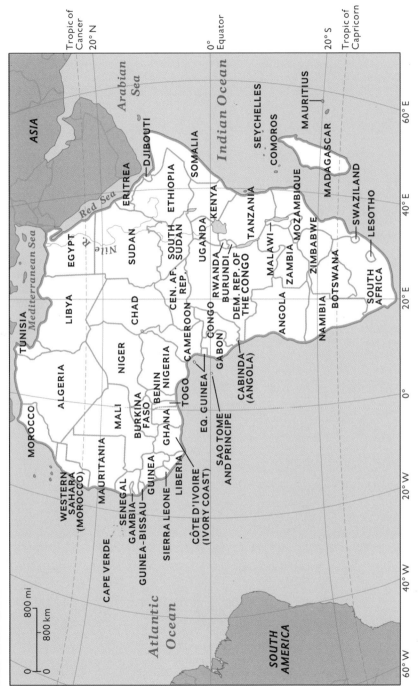

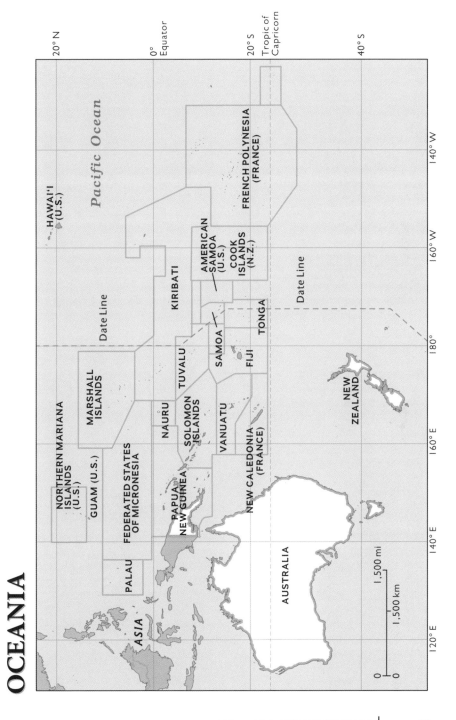

WORLD AND CONTINENT MAPS | PAGE 17

ANTARCTICA

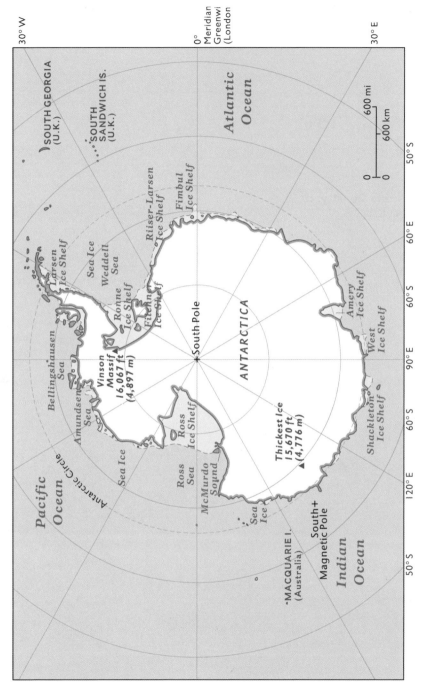

AFGHANISTAN

Country Name: Transitional Islamic State of Afghanistan
Continent: Asia
Area: 251,773 sq mi / 652,090 sq km
Population: 32,358,000
Population Density: 129 people per sq mi / 50 people per sq km
Capital: Kabul

PHYSICAL:

Highest Point: Nowshak 24,557 ft / 7,485 m
Lowest Point: Amu Darya (river) 846 ft / 258 m
Mountain Ranges:
   Hindu Kush: •cover most of Afghanistan •extend into Tajikistan, Pakistan, and China •include the Safed Koh Range
Peaks (minimum elevation of 8,000 ft / 2,400 m):
   Nowshak: •located in the Hindu Kush •highest point in Afghanistan
Passes:
   Khyber: •located in the Safed Koh Range •crosses the border into Pakistan

Rivers:
   Amu Darya: •forms all of the Afghanistan-Uzbekistan border •forms part of Afghanistan's border with Turkmenistan and Tajikistan
   Helmand: •has its source in the Hindu Kush mountains •extends into Iran

POLITICAL:
Independence: August 19, 1919 (from the United Kingdom)
Former Name: Republic of Afghanistan
Bordering Countries: Pakistan, Tajikistan, Iran, Turkmenistan, Uzbekistan, China (6)
Regions:
   Rigestan: •arid desert region in southern Afghanistan
Administrative Divisions: Badakhshan, Badghis, Baghlan, Balkh, Bamian, Daykundi, Farah, Faryab, Ghazni, Ghowr, Helmand, Herat, Jowzjan, Kabol, Kandahar, Kapisa, Khost, Konar, Kondoz, Laghman, Lowgar, Nangarhar, Nimruz, Nuristan, Oruzgan, Paktia, Paktika, Panishir, Parvan, Samangan, Sar-e Pol, Takhar, Vardak, Zabol (34 provinces)
Ethnic/Racial Groups: Pashtun, Tajik, Hazara, Uzbek
Religion: Islam (primarily Sunni and Shiite)
Languages: Pashtu, Afghan Persian (Dari), Uzbek, Turkmen
Currency: afghani
Current President: Hamid Karzai
Cities (capital, largest, or with at least a million people):
   Kabul: •located in eastern Afghanistan •city in the Kabol province •capital of Afghanistan •most populated city in Afghanistan (3,573,000 people) •located in the Hindu Kush

ENVIRONMENTAL/ECONOMIC:
Climate: arid to semi-arid with cold winters/hot summers
Natural Resources: natural gas, petroleum, coal, copper, chromite
Agricultural Products: opium, wheat, fruits and nuts, wool
Major Exports: opium, fruits and nuts, handwoven carpets, wool, cotton
Natural Hazards: earthquakes, floods, droughts

ALBANIA
Country Name: Republic of Albania
Continent: Europe
Area: 11,100 sq mi / 28,748 sq km
Population: 3,197,000
Population Density: 288 people per sq mi / 111 people per sq km
Capital: Tirana

PHYSICAL:
Highest Point: Golem Korab 9,068 ft / 2,764 m
Lowest Point: Adriatic Sea 0 ft / 0 m (sea level)

Mountain Ranges:
  Pindus: •located in southern Albania •extend into Greece
Peaks (minimum elevation of 8,000 ft / 2,400 m):
  Golem Korab: •on the Albania-Macedonia border •highest point in Albania
Seas:
  Adriatic: •forms part of the western coast of Albania •feeds into the Ionian Sea
  Ionian: •forms part of the southwestern coast of Albania •feeds into the Mediterranean Sea
Straits:
  Strait of Otranto: •forms part of the southwestern coast of Albania •separates Albania from Italy •connects the Adriatic and Ionian Seas
Lakes:
  Ohrid: •located in eastern Albania •on the Albania-Macedonia border
  Prespa: •located in eastern Albania •on the border with Macedonia and Greece
  Scutari: •located in northwestern Albania •on the Albania–Montenegro border

POLITICAL:

Independence: November 28, 1912 (from the Ottoman Empire)
Former Name: People's Socialist Republic of Albania
Bordering Countries: Montenegro, Greece, Macedonia, Kosovo (4)
Administrative Divisions: Beratit, Dibres, Durresit, Elbasanit, Fierit, Gjirokastres, Korces, Kukesit, Lezhes, Shkodres, Tirane, Vlores (12 counties)

Ethnic/Racial Groups: Albanian, Greek
Religions: Islam, Christianity (Albanian Orthodox, Roman Catholic)
Languages: Albanian (Tosk), Greek
Currency: lek
Current President: Bamir Topi
Cities (capital, largest, or with at least a million people):
   Tirana: •located in central Albania •city in Tirane county •capital of Albania •most populated city in Albania (433,000 people)

ENVIRONMENTAL/ECONOMIC:
Climate: mild temperature, cool and wet winters/hot and dry summers
Natural Resources: oil, natural gas, coal, chromium, copper
Agricultural Products: wheat, corn, potatoes, vegetables, meat
Major Exports: textiles and footwear, asphalt, metals and metallic ores, crude oil
Natural Hazards: earthquakes, tsunamis, floods, droughts

ALGERIA
Country Name: People's Democratic Republic of Algeria
Continent: Africa
Area: 919,595 sq mi / 2,381,741 sq km (largest in Africa)
Population: 35,980,000
Population Density: 39 people per sq mi / 15 people per sq km
Capital: Algiers

PHYSICAL:
Highest Point: Tahat 9,582 ft / 3,003 m
Lowest Point: Chott Melrhir 131 ft / 40 m below sea level
Mountain Ranges:
  Ahaggar: •located in southern Algeria
  Atlas: •located in northern Algeria •extend into Morocco and Tunisia
Peaks (minimum elevation of 8,000 ft / 2,400 m):
  Tahat: •located in the Ahaggar Mountains •highest point in Algeria
Plateau Regions:
  Hauts Plateaux: •located in northern Algeria •extend into Morocco •high basin in the Atlas Mountains
Depressions:
  Chott Melrhir: •located near the Grand Erg Oriental •lowest point in Algeria
Deserts:
  Grand Erg Occidental: •located in central Algeria
  Grand Erg Oriental: •located in eastern Algeria •extends into Tunisia
  Sahara: •covers most of Algeria •extends into Western Sahara (claimed by Morocco), Mauritania, Mali, Niger, Libya, and Tunisia •largest desert in the world
  Tenere: •located in southern Algeria •extends into Niger
Seas:
  Alboran (sometimes labeled as part of the Mediterranean): •forms part of the northwestern coast of Algeria •feeds into the Atlantic Ocean
  Mediterranean: •forms the northern coast of Algeria •feeds into the Atlantic Ocean

POLITICAL:
Independence: July 5, 1962 (from France)
Bordering Countries: Morocco, Mali, Libya, Tunisia, Niger, Mauritania, Western Sahara (claimed by Morocco) (7)
Administrative Divisions: Adrar, Ain Defla, Ain Temouchent, Alger, Annaba, Batna, Bechar, Bejaia, Biskra, Blida, Bordj Bou Arreridj, Bouira, Boumerdes, Chlef, Constantine, Djelfa, El Bayadh, El Oued, El Tarf, Ghardaia, Guelma, Illizi, Jijel, Khenchela, Laghouat, Mascara, Medea, Mila, Mostaganem, M'Sila, Naama, Oran, Ouargla, Oum el Bouaghi, Relizane, Saida, Setif, Sidi Bel Abbes, Skikda, Souk Ahras, Tamanghasset, Tebessa, Tiaret, Tindouf, Tipaza, Tissemsilt, Tizi Ouzou, Tlemcen (48 provinces)
Ethnic/Racial Groups: Arab-Berber
Religion: Islam (predominantly Sunni)
Languages: Arabic, French, Berber dialects
Currency: Algerian dinar
Current President: Abdelaziz Bouteflika
Cities (capital, largest, or with at least a million people):
  Algiers: •located in northern Algeria •city in the Alger province •capital of Algeria •most populated city in Algeria (2,740,000 people) •chief port on the Mediterranean

ENVIRONMENTAL/ECONOMIC:
Climate: arid to semi-arid with mild and wet winters/hot and dry summers along the coast, hot and dry in the desert
Natural Resources: oil, natural gas, iron ore, phosphates, uranium
Agricultural Products: wheat, barley, oats, grapes, sheep
Major Exports: petroleum, natural gas, petroleum products
Natural Hazards: earthquakes, mudslides, floods, droughts, desertification

ANDORRA
Country Name: Principality of Andorra
Continent: Europe
Area: 181 sq mi / 468 sq km
Population: 85,000
Population Density: 470 people per sq mi / 182 people per sq km
Capital: Andorra

PHYSICAL:
Highest Point: Coma Pedrosa 9,666 ft / 2,946 m
Lowest Point: Riu Runer 2,756 ft / 840 m
Mountain Ranges:
   Pyrenees: •cover the entire country •extend into France and Spain
Peaks (minimum elevation of 8,000 ft / 2,400 m):
   Coma Pedrosa: •located in the Pyrenees •on the Andorra-Spain border

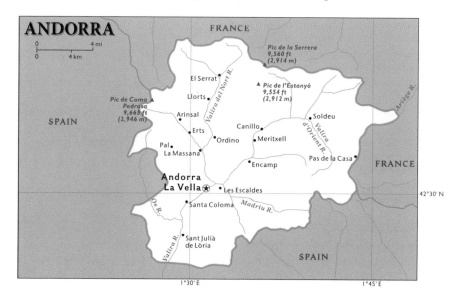

POLITICAL:
Independence: 1278 (formed under joint sovereignty of France and Spain)
Bordering Countries: Spain, France (2)
Administrative Divisions: Andorra, Canillo, Encamp, La Massana, Escaldes-Engordany, Ordino, Sant Julia de Loria (7 parishes)
Ethnic/Racial Groups: Spanish, Andorran, Portuguese, French
Religion: Christianity (predominantly Roman Catholic)
Languages: Catalan, French, Castilian, Portuguese
Currency: euro
Heads of State: President of France and Bishop of Urgel (Spain)
Current Head of Government: Marc Forne Molne
Cities (capital, largest, or with at least a million people):
  Andorra La Vella: • located in western Andorra • city in the Andorra la Vella parish • capital of Andorra • most populated city in Andorra (25,000 people) • located in the Pyrenees

ENVIRONMENTAL/ECONOMIC:
Climate: temperate with snowy and cold winters/warm and dry summers
Natural Resources: hydropower, mineral water, timber, iron ore, lead
Agricultural Products: rye, wheat, barley, oats, sheep
Major Exports: tobacco products, furniture
Natural Hazards: avalanches

ANGOLA
Country Name: Republic of Angola
Continent: Africa
Area: 481,354 sq mi / 1,246,700 sq km
Population: 19,638,000
Population Density: 41 people per sq mi / 16 people per sq km
Capital: Luanda

PHYSICAL:
Highest Point: Morro de Moco 8,596 ft / 2,620 m
Lowest Point: Atlantic Ocean 0 ft / 0 m (sea level)
Peaks (minimum elevation of 8,000 ft / 2,400 m):
  Morro de Moco: • located near the Bie Plateau • highest point in Angola
Plateau Regions:
  Bie Plateau: • located in central Angola
Deserts:
  Namib: • located in southwestern Angola • extends into Namibia
Oceans:
  Atlantic: • forms the western coast of Angola

Rivers:
    Congo: •has its mouth in the Atlantic Ocean •forms part of the Angola–Democratic Republic of the Congo border •extends into Democratic Republic of the Congo
    Cuango: •has its source in the Bie Plateau •forms part of the Angola–Democratic Republic of the Congo border •extends into Democratic Republic of the Congo
    Cubango: •has its source in the Bie Plateau •forms part of the Angola-Namibia border •extends into Namibia
    Zambezi: •extends into Zambia

Deltas:
    Congo River Delta: •mouth of the Congo River •feeds into the Atlantic Ocean •borders Angola and Democratic Republic of the Congo

POLITICAL:
Independence: November 11, 1975 (from Portugal)
Former Name: People's Republic of Angola
Bordering Countries: Democratic Republic of the Congo, Namibia, Zambia, Republic of the Congo (4)
Administrative Divisions: Bengo, Benguela, Bie, Cabinda, Cuando Cubango, Cuanza Norte, Cuanza Sul, Cunene, Huambo, Huila, Luanda, Lunda Norte, Lunda Sul, Malanje, Moxico, Namibe, Uige, Zaire (18 provinces)
Ethnic/Racial Groups: Ovimbundu, Kimbundu, Bakongo
Religions: indigenous beliefs, Christianity (Roman Catholic, Protestant)
Languages: Portuguese, Bantu
Currency: kwanza

Current President: Jose Eduardo dos Santos
Cities (capital, largest, or with at least a million people):
Luanda: •located in northwestern Angola •city in the Luanda province •capital of Angola •most populated city in Angola (4,511,000 people) •chief port on the Atlantic Ocean

ENVIRONMENTAL/ECONOMIC:
Climate: semi-arid in the south and along the coast with cool summers/hot winters
Natural Resources: oil, diamonds, iron ore, phosphates, copper
Agricultural Products: bananas, sugarcane, coffee, sisal, livestock
Major Exports: crude oil, diamonds, refined petroleum products, natural gas
Natural Hazards: flooding

ANTIGUA AND BARBUDA
Country Name: Antigua and Barbuda
Continent: North America
Area: 171 sq mi / 442 sq km
Population: 88,000
Population Density: 515 people per sq mi / 199 people per sq km
Capital: Saint John's

PHYSICAL:
Highest Point: Boggy Peak 1,319 ft / 402 m
Lowest Point: Caribbean Sea 0 ft / 0 m (sea level)
Oceans:
Atlantic: •forms the entire coast of Antigua and Barbuda
Islands:
Antigua: •largest island in Antigua and Barbuda •surrounded by the Atlantic Ocean
Barbuda: •island in Antigua and Barbuda •surrounded by the Atlantic Ocean

POLITICAL:
Independence: November 1, 1981 (from the United Kingdom)
Administrative Divisions: Saint George, Saint John, Saint Mary, Saint Paul, Saint Peter, Saint Philip (6 parishes) •Barbuda, Redonda (2 dependencies)
Ethnic/Racial Groups: black, British, Portuguese, Lebanese, Syrian
Religion: Christianity (predominantly Protestant with some Roman Catholic)
Languages: English, local dialects
Currency: East Caribbean dollar
Current Prime Minister: Baldwin Spencer
Cities (capital, largest, or with at least a million people):
Saint John's: •located on the island of Antigua •city in the Saint John parish •capital of Antigua and Barbuda •most populated city in Antigua and Barbuda (27,000 people) •chief port on the Atlantic Ocean

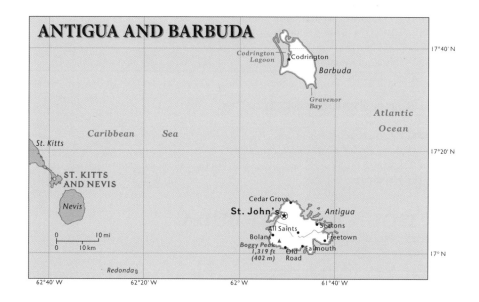

ENVIRONMENTAL/ECONOMIC:
Climate: tropical marine, little change in temperature
Agricultural Products: cotton, fruit, vegetables, bananas, livestock
Major Exports: petroleum products, manufactures, machinery and transport equipment, food, and live animals
Natural Hazards: hurricanes, tropical storms, droughts

ARGENTINA
Country Name: Argentine Republic
Continent: South America
Area: 1,073,518 sq mi / 2,780,400 sq km
Population: 40,488,000
Population Density: 38 people per sq mi / 15 people per sq km
Capital: Buenos Aires

PHYSICAL:
Highest Point: Cerro Aconcagua 22,834 ft / 6,960 m
Lowest Point: Laguna del Carbon 344 ft / 105 m below sea level
Mountain Ranges:
  Andes: •located in western Argentina •extend into Chile and Bolivia
  Sierra de Cordoba: •located in central Argentina

Peaks (minimum elevation of 8,000 ft / 2,400 m):
   Cerro Aconcagua: •located in the Andes   •highest point in South America
Grasslands/Prairies:
   Gran Chaco: •located in northern Argentina   •extends into Paraguay
   Pampas: •located in central Argentina
Depressions:
   Valdes Peninsula: •located on the southeastern coast of Argentina   •lowest point in South America
Oceans:
   Atlantic: •forms the eastern coast of Argentina
Islands:
   Islas Malvinas: •group of islands off the southeast coast   •claimed by Argentina   •administered by the United Kingdom as the Falkland Islands
   Tierra del Fuego: •group of islands south of mainland Argentina   •name of the largest island in the group   •bordered by the Atlantic Ocean   •western half belongs to Chile
Straits:
   Strait of Magellan: •forms part of the southern coast of Argentina   •separates Tierra del Fuego from mainland Argentina   •connects the Pacific and Atlantic Oceans
Rivers:
   Paraguay: •tributary of the Parana River   •forms part of the Argentina-Paraguay border   •extends into Brazil
   Parana: •has its mouth in the Rio de la Plata Estuary   •forms part of the Argentina-Paraguay border   •extends into Brazil

Uruguay: •has its mouth in the Rio de la Plata Estuary •forms all of the Argentina-Uruguay border •extends into Brazil

Deltas:
Rio de la Plata Estuary: •mouth of the Parana and Uruguay Rivers •feeds into the Atlantic Ocean •borders Argentina and Uruguay

Waterfalls:
Iguazu Falls: •located in northeastern Argentina •part of the Iguazu River •on the Argentina-Brazil border

POLITICAL:
Independence: July 9, 1816 (from Spain)
Bordering Countries: Chile, Paraguay, Brazil, Bolivia, Uruguay (5)
Regions:
Patagonia: •high plateau in southern Argentina •located east of the Andes
Administrative Divisions: Buenos Aires, Catamarca, Chaco, Chubut, Cordoba, Corrientes, Entre Rios, Formosa, Jujuy, La Pampa, La Rioja, Mendoza, Misiones, Neuquen, Rio Negro, Salta, San Juan, San Luis, Santa Cruz, Santa Fe, Santiago del Estero, Tierra del Fuego, Tucuman (23 provinces) •Buenos Aires Federal District (1 federal district)
Ethnic/Racial Groups: European (mainly Spanish and Italian), Amerindian
Religion: Christianity (predominantly Roman Catholic)
Languages: Spanish, English, Italian, German, French
Currency: Argentine peso
Current President: Cristina Fernandez de Kirchner
Cities (capital, largest, or with at least a million people):
Buenos Aires: •located in eastern Argentina •city in the Buenos Aires province •Buenos Aires Federal District is located in center of city •capital of Argentina •most populated city in Argentina (12,988,000 people) •chief port on the Rio de la Plata Estuary
Cordoba: •located in central Argentina •city in the Cordoba province •on the edge of the Sierra de Cordoba
Rosario: •located in central Argentina •city in the Santa Fe province •chief port on the Parana River
Mendoza: •located in western Argentina •city in the Mendoza province •on the edge of the Andes

ENVIRONMENTAL/ECONOMIC:
Climate: mostly temperate, arid in the south
Natural Resources: lead, zinc, tin, copper, iron ore
Agricultural Products: sunflower seeds, lemons, soybeans, livestock
Major Exports: edible oils, fuels and energy, cereals, feed, motor vehicles
Natural Hazards: earthquakes, pamperos (windstorms), floods

ARMENIA

Country Name: Republic of Armenia
Continent: Asia
Area: 11,484 sq mi / 29,743 sq km
Population: 3,123,000
Population Density: 272 people per sq mi / 105 people per sq km
Capital: Yerevan

PHYSICAL:
Highest Point: Aragats 13,419 ft / 4,090 m
Lowest Point: Debed (river) 1,312 ft / 400 m
Mountain Ranges:
   Lesser Caucasus: •cover most of the country •extend into Georgia, Turkey, and Azerbaijan
Peaks (minimum elevation of 8,000 ft / 2,400 m):
   Aragats: •located in the Lesser Caucasus •highest point in Armenia
Lakes:
   Sevan: •located in central Armenia •surrounded by the Lesser Caucasus
Rivers:
   Aras: •forms all of the Armenia-Iran border •forms part of the Armenia-Turkey border •extends into Turkey and Azerbaijan

POLITICAL:
Independence: September 21, 1991 (from the Soviet Union)
Former Names: Armenian Soviet Socialist Republic, Armenian Republic
Bordering Countries: Azerbaijan, Turkey, Georgia, Iran (4)
Administrative Divisions: Aragatsotn, Ararat, Armavir, Gegharkunik, Kotayk, Lorri, Shirak, Syunik, Tavush, Vayots Dzor, Yerevan (11 provinces)
Ethnic/Racial Groups: Armenian, Russian
Religion: Christianity (primarily Armenian Apostolic)
Languages: Armenian, Russian
Currency: dram
Current President: Serzh Sargsian
Cities (capital, largest, or with at least a million people):
    Yerevan: •located in central Armenia •city in the Yerevan province •capital of Armenia •most populated city in Armenia (1,110,000 people)

ENVIRONMENTAL/ECONOMIC:
Climate: Highland continental with hot summers/cold winters
Natural Resources: gold, copper, molybdenum, zinc, alumina
Agricultural Products: fruits (especially grapes), vegetables, livestock
Major Exports: diamonds, mineral products, foodstuffs
Natural Hazards: earthquakes, droughts

AUSTRALIA
Country Name: Commonwealth of Australia
Continent: Australia/Oceania
Area: 2,969,906 sq mi / 7,692,024 sq km (largest in Australia/Oceania)
Population: 22,670,000 (largest in Australia/Oceania)
Population Density: 8 people per sq mi / 3 people per sq km (least dense in Australia/Oceania)
Capital: Canberra

PHYSICAL:
Highest Point: Mount Kosciuszko 7,310 ft / 2,228 m
Lowest Point: Lake Eyre 52 ft / 16 m below sea level (lowest point in Australia/Oceania)
Mountain Ranges:
    Australian Alps: •located in southeastern Australia
    Flinders Ranges: •located in southern Australia
    Great Dividing Range: •located in eastern Australia
    Macdonnell Ranges: •located in central Australia
Depressions:
    Lake Eyre: •located in central Australia •lowest point in Australia/Oceania

Grasslands/Prairies:
  Nullarbor Plain: •located in southern Australia •north of the Great Australian Bight
Plateau Regions:
  Kimberley Plateau: •located in northwestern Australia
  Lake Eyre Basin: •located in central Australia
  Murray River Basin: •located in southern Australian •between the Flinders Ranges and the Great Dividing Range
Deserts:
  Gibson: •located in western Australia
  Great Sandy: •located in western Australia
  Great Victoria: •located in southern Australia
  Simpson: •located in central Australia
  Tanami: •located in northern Australia
Oceans:
  Indian: •forms the western and southern coasts of Australia •forms part of the northern coast of Australia
  Pacific: •forms part of the eastern coast of Australia
Seas:
  Arafura Sea: •forms part of the northern coast of Australia
  Coral Sea: •forms part of the eastern coast of Australia
  Tasman Sea: •forms part of the southeastern coast of Australia
  Timor Sea: •forms part of the northern coast of Australia

Gulfs:
>   Gulf of Carpentaria: •forms part of the northern coast of Australia •feeds into the Arafura Sea
>   Joseph Bonaparte: •forms part of the northern coast of Australia •feeds into the Timor Sea
>   Spencer: •forms part of the southern coast of Australia

Bays:
>   Great Australian Bight: •forms part of the southern coast of Australia •feeds into the Indian Ocean

Straits:
>   Bass: •forms part of the southwestern coast of Australia •separates the island of Tasmania from mainland Australia •connects the Indian Ocean and the Tasman Sea
>   Torres: •forms part of the northern coast of Australia •separates Australia from Papua New Guinea •connects the Arafura and Coral Seas

Lakes:
>   Eyre: •located in southern Australia •located on the Lake Eyre Basin •lowest point in Australia/Oceania •largest lake in Australia/Oceania
>   Torrens: •located in southern Australia

Rivers:
>   Darling: •tributary of the Murray River • has its source in the Great Dividing Range •along with the Murray River, it forms the longest river in Australia/Oceania
>   Murray: •has its mouth in the Indian Ocean •has its source in the Australian Alps •including its tributary the Darling River, the Murray forms the longest river in Australia/Oceania

Peninsulas:
>   Cape York: •bordered by the Coral Sea, Gulf of Carpentaria, and Torres Strait
>   Eyre: •bordered by the Indian Ocean, Great Australian Bight, and Spencer Gulf

Islands:
>   Tasmania: •an island southeast of mainland Australia •bordered by the Indian Ocean, Tasman Sea, and Bass Strait •a state of Australia

POLITICAL:

Independence: January 1, 1901 (from the United Kingdom)

Regions:
>   Arnhem Land: •located in Australia's Northern Territory •Aboriginal homeland

Administrative Divisions: New South Wales, Queensland, South Australia, Tasmania, Victoria, Western Australia (6 states) •Australian Capital Territory, Northern Territory (2 territories)

External Territories: Ashmore and Cartier Islands, Christmas Island, Cocos Islands, Coral Sea Islands, Heard and McDonald Islands, Norfolk Island (6)

Ethnic/Racial Groups: white, Asian, Aboriginal

Religion: Christianity (Anglican, Roman Catholic)

Languages: English, indigenous languages

Currency: Australian dollar
Current Prime Minister: Julia Gillard
Cities (capital, largest, or with at least a million people):
- Sydney: •located in southeastern Australia •city in the New South Wales state •most populated city in Australia (4,429,000 people) •chief port on the Tasman Sea
- Melbourne: •located in southeastern Australia •city in the Victoria state •chief port on the Bass Strait
- Brisbane: •located in eastern Australia •city in the Queensland state •chief port on the Coral Sea
- Perth: •located in southwestern Australia •city in the Western Australia state •a major port on the Indian Ocean
- Adelaide: •located in southern Australia •city in the South Australia state •a major port on the Indian Ocean
- Canberra: •city in the Australian Capital Territory •capital of Australia (384,000 people)

ENVIRONMENTAL/ECONOMIC:
Climate: mostly arid to semi-arid, temperate in the south and east, tropical in the north
Natural Resources: bauxite, coal, iron ore, copper, tin
Agricultural Products: wheat, barley, sugarcane, fruits, cattle
Major Exports: coal, gold, meat, wool, alumina
Natural Hazards: cyclones, severe droughts, forest fires

AUSTRIA
Country Name: Republic of Austria
Continent: Europe
Area: 32,378 sq mi / 83,858 sq km
Population: 8,418,000
Population Density: 260 people per sq mi / 100 people per sq km
Capital: Vienna

PHYSICAL:
Highest Point: Grossglockner 12,461 ft / 3,798 m
Lowest Point: Neusiedler See 377 ft / 115 m
Mountain Ranges:
- Alps: •cover most of southern Austria •extend into Switzerland, Liechtenstein, Italy, and Germany

Peaks (minimum elevation of 8,000 ft / 2,400 m):
- Grossglockner: •located in the Alps •highest point in Austria

Passes:
- Brenner: •located in the Alps •crosses the border into Italy

Lakes:
- Constance: •located in northwestern Austria •on the border with Germany and Switzerland •fed by the Rhine River
- Neusiedler See: •located in eastern Austria •on the Austria-Hungary border

Rivers:
> Danube: •forms part of the Austria-Germany border •extends into Germany, Slovakia, and Hungary
> Drava: •has its source in the Alps •extends into Slovenia
> Inn: •tributary of the Danube •forms part of the Austria-Germany border •has its source in the Swiss Alps
> Rhine: •forms part of the Austria-Liechtenstein border •extends into Switzerland and Germany

POLITICAL:
Independence: 1156 (from Bavaria)
Bordering Countries: Germany, Italy, Hungary, Czech Republic, Slovenia, Switzerland, Slovakia, Liechtenstein (8)
Administrative Divisions: Burgenland, Carinthia, Lower Austria, Salzburg, Styria, Tirol, Upper Austria, Vienna, Vorarlberg (9 states)
Ethnic/Racial Groups: primarily German
Religion: Christianity (Roman Catholic, Protestant)
Languages: German, Slovene, Croatian, Hungarian
Currency: euro
Current President: Heinz Fischer
Cities (capital, largest, or with at least a million people):
> Vienna: •located in northeastern Austria •city in the Vienna state •capital of Austria •most populated city in Austria (1,693,000 people) •chief port on the Danube River

ENVIRONMENTAL/ECONOMIC:
Climate: temperate with cold and rainy or snowy winters/moderate summers
Natural Resources: iron ore, oil, timber, magnesite, lead
Agricultural Products: grains, potatoes, sugar beets, wine, dairy products
Major Exports: machinery and equipment, motor vehicles and parts, paper, metal goods, chemicals
Natural Hazards: landslides, avalanches, earthquakes

AZERBAIJAN
Country Name: Republic of Azerbaijan
Continent: Asia
Area: 33,436 sq mi / 86,600 sq km
Population: 9,150,000
Population Density: 274 people per sq mi / 106 people per sq mi
Capital: Baku

PHYSICAL:
Highest Point: Bazar Dyuzi 14,652 ft / 4,466 m
Lowest Point: Caspian Sea 92 ft / 28 m below sea level
Mountain Ranges:
  Caucasus: •located in northern Azerbaijan •extend into Russia and Georgia
  Lesser Caucasus: •located in western Azerbaijan •extend into Armenia and Georgia

Peaks (minimum elevation of 8,000 ft / 2,400 m):
  Bazar Dyuzi: •located in the Caucasus •on the Azerbaijan-Russia border •highest point in Azerbaijan
Depressions:
  Caspian Sea: •located in eastern Azerbaijan •lowest point in Azerbaijan
Lakes:
  Caspian Sea: •forms the eastern coast of Azerbaijan •also borders Russia and Iran •largest lake in the world
Rivers:
  Aras: •has its mouth in the Caspian Sea •forms part of the Azerbaijan-Iran border •extends into Turkey

POLITICAL:
Independence: August 30, 1991 (from the Soviet Union)
Former Name: Azerbaijan Soviet Socialist Republic
Bordering Countries: Armenia, Iran, Georgia, Russia, Turkey (5)
Regions:
  Nagorno-Karabakh: •mountainous region in western Azerbaijan •located in the Lesser Caucasus
Administrative Divisions: Abseron, Agcabadi, Agdam, Agdas, Agstafa, Agsu, Astara, Balakan, Barda, Beylaqan, Bilasuvar, Cabrayil, Calilabad, Daskasan, Davaci, Fuzuli, Gadabay, Goranboy, Goycay, Haciqabul, Imisli, Ismayilli, Kalbacar, Kurdamir, Lacin, Lankaran, Lerik, Masalli, Neftcala, Oguz, Qabala, Qax, Qazax, Qobustan, Quba, Qubadli, Qusar, Saatli, Sabirabad, Saki, Salyan, Samaxi, Samkir, Samux, Siyazan, Susa, Tartar, Tovuz, Ucar, Xacmaz, Xanlar, Xizi, Xocali, Xocavand, Yardimli, Yevlax, Zangilan, Zaqatala, Zardab (59 rayons) •Ali Bayramli, Baki, Ganca, Lankaran, Mingacevir, Naftalan, Saki, Sumqayit, Susa, Xankandi, Yevlax (11 cities) •Naxcivan (1 autonomous republic)
Ethnic/Racial Groups: Azeri, Dagestani, Russian, Armenian
Religion: primarily Islam
Languages: Azeri, Russian, Armenian
Currency: Azerbaijani manat
Current President: Ilham Aliyev
Cities (capital, largest, or with at least a million people):
  Baku: •located in eastern Azerbaijan •capital of Azerbaijan •most populated city in Azerbaijan (1,950,000 people) •chief port on the Caspian Sea

ENVIRONMENTAL/ECONOMIC:
Climate: dry, semi-arid steppe
Natural Resources: oil, natural gas, iron ore, nonferrous metals, alumina
Agricultural Products: cotton, grain, rice, grapes, cattle
Major Exports: oil and gas, machinery, cotton, foodstuffs
Natural Hazards: droughts

BAHAMAS

Country Name: Commonwealth of the Bahamas
Continent: North America
Area: 5,382 sq mi / 13,939 sq km
Population: 357,000
Population Density: 66 people per sq mi / 26 people per sq km
Capital: Nassau

PHYSICAL:
Highest Point: Mount Alvernia 207 ft / 63 m
Lowest Point: Atlantic Ocean 0 ft / 0 m (sea level)
Oceans:
  Atlantic: •surrounds the islands of the Bahamas
Islands:
  Andros: •largest island in the Bahamas •surrounded by the Atlantic Ocean
  Great Inagua: •second largest island in the Bahamas •surrounded by the Atlantic Ocean

POLITICAL:
Independence: July 10, 1973 (from the United Kingdom)
Administrative Divisions: Acklins Islands, Berry Islands, Bimini, Black Point, Cat Island, Central Abaco, Central Andros, Central Eleuthera, City of Freeport, Crooked Island and Long Key, East Grand Bahama, Exuma, Grand Key, Harbour Island, Hope Town, Inagua, Long Island, Mangrove Key, Mayaguana, Moore's Island, North Abaco, North Andros, North Eleuthera, Ragged Island, Rum Cay, San Salvador, South Abaco, South Andros, South Eleuthera, Spanish Wells, West Grand Bahama (31 districts)
Ethnic Groups: black, white
Religion: Christianity (Baptist, Anglican, Roman Catholic)
Languages: English, Creole
Currency: Bahamian dollar
Current Prime Minister: Hubert Ingraham
Cities (capital, largest, or with at least a million people):
  Nassau: •located on New Providence island •city in the New Providence district •capital of the Bahamas •most populated city in the Bahamas (248,000 people) •chief port on the Atlantic Ocean

ENVIRONMENTAL/ECONOMIC:
Climate: tropical marine, moderated by the Gulf Stream current
Natural Resources: salt, aragonite, timber, arable land

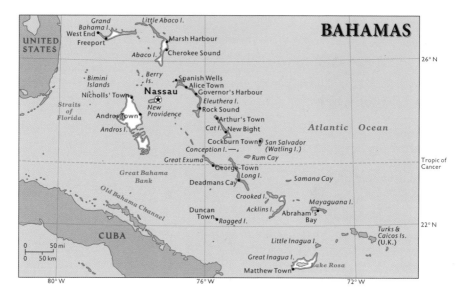

Agricultural Products: citrus, vegetables, poultry
Major Exports: fish and crawfish, rum, salt, chemicals, fruits and vegetables
Natural Hazards: hurricanes, tropical storms

BAHRAIN
Country Name: Kingdom of Bahrain
Continent: Asia
Area: 277 sq mi / 717 sq km
Population: 1,336,000
Population Density: 4,823 people per sq mi / 1,925 people per sq km
Capital: Manama

PHYSICAL:
Highest Point: Jabal ad Dukhan 400 ft / 122 m
Lowest Point: Persian Gulf 0 ft / 0 m (sea level)
Gulfs:
    Persian: •forms the northern coast of Bahrain
    Gulf of Bahrain: •forms the western, southern, and eastern coasts of Bahrain
Islands:
    Bahrain: •largest island in Bahrain •surrounded by Persian Gulf and Gulf of Bahrai

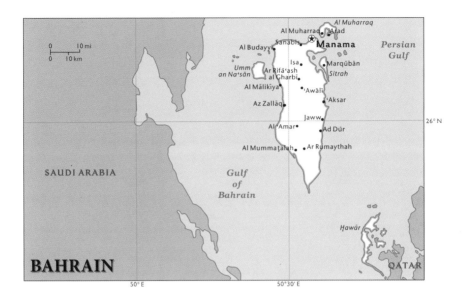

POLITICAL:
Independence: August 15, 1971 (from the United Kingdom)
Former Name: Dilmun
Administrative Divisions: Al Asimah, Al Janubiyah, Al Muharraq, Ash Shamaliyah, Al Wusta (5 governorates)
Ethnic/Racial Groups: Arab, Asian, Iranian
Religion: Islam (Sunni, Shiite)
Languages: Arabic, English, Farsi, Urdu
Currency: Bahraini dinar
Current King: Hamad bin Isa Al Khalifa
Cities (capital, largest, or with at least a million people):
  Manama: •located on the island of Bahrain •city in the Al Manamah municipality •capital of Bahrain •most populated city in Bahrain (163,000 people) •chief port on the Persian Gulf

ENVIRONMENTAL/ECONOMIC:
Climate: arid with mild winters/hot summers
Natural Resources: petroleum, natural gas, fish, pearls
Agricultural Products: fruits, vegetables, poultry, shrimp
Major Exports: petroleum and petroleum products, aluminum, textiles
Natural Hazards: droughts, dust storms

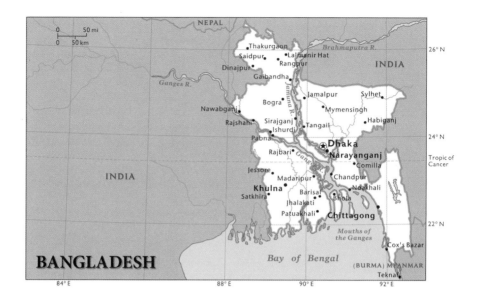

BANGLADESH

Country Name: People's Republic of Bangladesh
Continent: Asia
Area: 56,997 sq mi / 147,570 sq km
Population: 150,685,000
Population Density: 2,644 people per sq mi / 1,021 people per sq km
Capital: Dhaka

PHYSICAL:
Highest Point: Keokradong 4,035 ft / 1,230 m
Lowest Point: Bay of Bengal 0 ft / 0 m (sea level)
Grasslands/Prairies:
　Sundarbans: •located in southern Bangladesh •borders the Bay of Bengal •extends into India
Bays:
　Bay of Bengal: •forms the southern coast of Bangladesh •feeds into the Indian Ocean
Rivers:
　Brahmaputra: •tributary of the Ganges •extends into India
　Ganges: •has its mouth in the Bay of Bengal •extends into India
　Jamuna: •tributary of the Ganges •joins with the Brahmaputra

Deltas:
    Mouths of the Ganges: •mouth of the Ganges •feed into the Bay of Bengal •extend into India

POLITICAL:
Independence: December 16, 1971 (from West Pakistan)
Former Name: East Pakistan
Bordering Countries: India, Myanmar (2)
Administrative Divisions: Barisal, Chittagong, Dhaka, Khulna, Rajshahi, Rangpur, Sylhet (7 divisions)
Ethnic/Racial Groups: Bengali
Religions: Islam, Hinduism
Languages: Bangla (Bengali), English
Currency: taka
Current Prime Minister: Sheikh Hasina Wajed
Cities (capital, largest, or with at least a million people):
    Dhaka: •located in central Bangladesh •city in the Dhaka division •capital of Bangladesh •most populated city in Bangladesh (14,251,000 people)
    Chittagong: •located in southeastern Bangladesh •city in the Chittagong division •chief port on the Bay of Bengal
    Khulna: •located in southwestern Bangladesh •city in the Khulna division •located near the Sundarbans in the Mouths of the Ganges

ENVIRONMENTAL/ECONOMIC:
Climate: tropical, mild winters/hot summers/warm and rainy monsoon seasons
Natural Resources: natural gas, arable land, timber, coal
Agricultural Products: rice, jute, tea, wheat, beef
Major Exports: garments, jute and jute goods, leather, frozen fish and seafood
Natural Hazards: droughts, cyclones, heavy rains, floods

BARBADOS
Country Name: Barbados
Continent: North America
Area: 166 sq mi / 430 sq km
Population: 274,000
Population Density: 1,651 people per sq mi / 637 people per sq km (most dense in North America)
Capital: Bridgetown

PHYSICAL:
Highest Point: Mount Hillaby 1,115 ft / 340 m
Lowest Point: Atlantic Ocean 0 ft / 0 m (sea level)

Oceans:
  Atlantic: •forms the entire coast of Barbados

POLITICAL:
Independence: November 30, 1966 (from the United Kingdom)
Administrative Divisions: Christ Church, Saint Andrew, Saint George, Saint James, Saint John, Saint Joseph, Saint Lucy, Saint Michael, Saint Peter, Saint Philip, Saint Thomas (11 parishes)
Ethnic/Racial Groups: predominantly black
Religion: Christianity (Protestant, Roman Catholic)
Language: English
Currency: Barbadian dollar
Current Prime Minister: Fruendel Stuart
Cities (capital, largest, or with at least a million people):
  Bridgetown: •located on the island of Barbados •city in the Saint Michael parish •capital of Barbados •most populated city in Barbados (112,000 people) •chief port on the Atlantic Ocean

ENVIRONMENTAL/ECONOMIC:
Climate: tropical
Natural Resources: oil, fish, natural gas
Agricultural Products: sugarcane, vegetables, cotton
Major Exports: sugar and molasses, rum, other foods and beverages, chemicals
Natural Hazards: hurricanes, landslides

BELARUS

Country Name: Republic of Belarus
Continent: Europe
Area: 80,153 sq mi / 207,595 sq km
Population: 9,472,000
Population Density: 118 people per sq mi / 46 people per sq km
Capital: Minsk

PHYSICAL:

Highest Point: Dzyarzhynskaya Hara 1,135 ft / 346 m
Lowest Point: Nyoman River 295 ft / 90 m
Swamps:
 Pinsk Marshes: •located in southern Belarus •extend into Ukraine
Rivers:
 Nyoman: •has its source in central Belarus •extends into Lithuania
 Dnyapro (Dnieper): •extends into Russia and Ukraine

POLITICAL:

Independence: August 25, 1991 (from the Soviet Union)
Former Name: Belorussian Soviet Socialist Republic
Bordering Countries: Russia, Ukraine, Lithuania, Poland, Latvia (5)
Administrative Divisions: Brestskaya, Homyelskaya, Hrodzyenskaya, Mahilyowskaya, Minskaya, Vitsyebskaya (6 voblasts) •Horad Minsk (1 municipality)

Ethnic/Racial Groups: Belarusian, Russian
Religions: Christianity (Eastern Orthodox, Roman Catholic, Protestant), Judaism, Islam
Languages: Belarusian, Russian
Currency: Belarusian ruble
Current President: Aleksander Lukashenko
Cities (capital, largest, or with at least a million people):
    Minsk: •located in central Belarus •city in the Horad Minsk municipality •capital of Belarus •most populated city in Belarus (1,837,000 people)

ENVIRONMENTAL/ECONOMIC:
Climate: continental/maritime, cold winters/cool summers
Natural Resources: forests, peat deposits, oil, natural gas, granite
Agricultural Products: grain, potatoes, vegetables, sugar beets, beef
Major Exports: machinery and equipment, mineral products, chemicals, metals, textiles
Natural Hazards: floods

BELGIUM
Country Name: Kingdom of Belgium
Continent: Europe
Area: 11,787 sq mi / 30,528 sq km
Population: 10,970,000
Population Density: 931 people per sq mi / 359 people per sq km
Capital: Brussels

PHYSICAL:
Highest Point: Signal de Botrange 2,277 ft / 694 m
Lowest Point: North Sea 0 ft / 0 m (sea level)
Seas:
    North: •forms the northwestern coast of Belgium •feeds into the Atlantic Ocean
Rivers:
    Meuse: •forms part of the Belgium-Netherlands border •extends into France and the Netherlands
    Schelde: •extends into France and the Netherlands
Regions:
    Ardennes: •forested plateau region in southeastern Belgium •extends into France and Luxembourg

POLITICAL:
Independence: October 4, 1830 (from the Netherlands)
Bordering Countries: France, Netherlands, Germany, Luxembourg (4)
Regions:

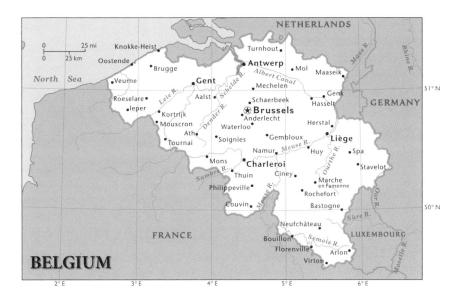

Flanders: •lowland region in northwestern Belgium •located east of the North Sea
Administrative Divisions: •Brussels-Capital, Flanders, Wallonia (3 regions)
Ethnic/Racial Groups: Fleming, Walloon
Religion: Christianity (Roman Catholic, Protestant)
Languages: Flemish (Dutch), French, German
Currency: euro
Current Prime Minister: Yves Leterme
Cities (capital, largest, or with at least a million people):
Brussels: •located in central Belgium •main city in the Brussels Capital Region •capital of Belgium •most populated city in Belgium (1,892,000 people)

ENVIRONMENTAL/ECONOMIC:
Climate: temperate with mild winters/cool summers
Natural Resources: coal, natural gas, construction materials, silica sand, carbonates
Agricultural Products: sugar beets, fresh vegetables, fruits, grain, beef
Major Exports: machinery and equipment, diamonds, metals and metal products
Natural Hazards: floods

BELIZE
Country Name: Belize
Continent: North America
Area: 8,867 sq mi / 22,965 sq km
Population: 318,000

Population Density: 36 people per sq mi / 14 people per sq km

Capital: Belmopan

PHYSICAL:
Highest Point: Doyles Delight 3,688 ft / 1,124 m
Lowest Point: Caribbean Sea 0 ft / 0 m (sea level)
Mountain Ranges:
  Maya: •located in southern Belize •extend into Guatemala
Seas:
  Caribbean: •forms most of the eastern coast of Belize •feeds into the Atlantic Ocean
Gulfs:
  Gulf of Honduras: •forms the southeastern coast of Belize •feeds into the Caribbean Sea
Bays:
  Bay of Chetumal: •forms part of the northern coast of Belize
Rivers:
  Hondo: •has its mouth in the Bay of Chetumal •feeds into the Caribbean Sea

POLITICAL:
Independence: September 21, 1981 (from the United Kingdom)
Former Name: British Honduras
Bordering Countries: Guatemala, Mexico (2)
Administrative Divisions: Belize, Cayo, Corozal, Orange Walk, Stann Creek, Toledo (6 districts)
Ethnic/Racial Groups: mestizo, Creole, Maya, Garifuna
Religion: Christianity (Roman Catholic, Protestant)

Languages: English, Spanish, Mayan, Garifuna, Creole
Currency: Belizean dollar
Current Prime Minister: Dean Oliver Barrow
Cities (capital, largest, or with at least a million people):
  Belize City: •located in eastern Belize •city in the Belize district •most populated city in Belize (56,000 people) •chief port on the Caribbean Sea
  Belmopan: •located in central Belize •city in the Cayo district •capital of Belize (20,000 people)

ENVIRONMENTAL/ECONOMIC:
Climate: tropical, hot and humid
Natural Resources: arable land potential, timber, fish, hydropower
Agricultural Products: bananas, coca, citrus, sugar, fish
Major Exports: sugar, bananas, citrus, clothing, fish products
Natural Hazards: hurricanes, floods

BENIN
Country Name: Republic of Benin
Continent: Africa
Area: 43,484 sq mi / 112,622 sq km
Population: 9,109,000
Population Density: 209 people per sq mi / 81 people per sq km
Capital: Porto-Novo

PHYSICAL:
Highest Point: Mont Sokbaro 2,159 ft / 658 m
Lowest Point: Bight of Benin 0 ft / 0 m (sea level)
Grasslands/Prairies:
  Sahel: •located in northern Benin •extends into Togo, Burkina Faso, Niger, and Nigeria
Bays:
  Bight of Benin: •forms the southern coast of Benin •feeds into the Gulf of Guinea
Rivers:
  Niger: •forms part of the Benin-Niger border •extends into Niger and Nigeria

POLITICAL:
Independence: August 1, 1960 (from France)
Former Name: Dahomey
Bordering Countries: Nigeria, Togo, Burkina Faso, Niger (4)
Administrative Divisions: Alibori, Atakora, Atlantique, Borgou, Collines, Kouffo, Donga, Littoral, Mono, Oueme, Plateau, Zou (12 departments)
Ethnic/Racial Groups: African (mainly Fon, Adja, Yoruba, Bariba)
Religions: indigenous beliefs, Christianity, Islam
Languages: French, Fon, Yoruba, tribal languages
Currency: Communaute Financiere Africaine franc
Current President: Thomas Yayi Boni
Cities (capital, largest, or with at least a million people):
  Cotonou: •located in southern Benin •city in the Littoral department •most populated city in Benin (815,000 people) •chief port on the Bight of Benin
  Porto-Novo: •located in southeastern Benin •city in the Oueme department •capital of Benin (276,000 people)

ENVIRONMENTAL/ECONOMIC:
Climate: tropical; hot and humid in the south; semi-arid in the north
Natural Resources: offshore oil deposits, limestone, marble, timber
Agricultural Products: cotton, corn, cassava, yams, livestock
Major Exports: cotton, crude oil, palm products, cacao
Natural Hazards: harmattan winds

BHUTAN
Country Name: Kingdom of Bhutan
Continent: Asia
Area: 17,954 sq mi / 46,500 sq km
Population: 708,000
Population Density: 39 people per sq mi / 15 people per sq km
Capital: Thimphu

PHYSICAL:
Highest Point: Gangkar Puensum 24,836 ft / 7,570 m
Lowest Point: Drangme Chhu 318 ft / 97 m
Mountain Ranges:
   Himalaya: •cover most of Bhutan •extend into India and China
Peaks (minimum elevation of 8,000 ft / 2,400 m):
   Gangkar Puensum: •located in the Himalaya •highest point in Bhutan

POLITICAL:
Independence: August 8, 1949 (from India)
Bordering Countries: India, China (2)
Administrative Divisions: Bumthang, Chhukha, Dagana, Gasa, Haa, Lhuentse, Mongar, Paro, Pemagatshel, Punakha, Samtse, Samdrup Jongkhar, Sharpang, Trashigang, Trashi Yangtse, Thimphu, Trongsa, Tsirang, Wangdu Phodrang, Zemgang (20 districts)
Ethnic/Racial Groups: Bhote, Nepalese, indigenous tribes
Religions: Lamaistic Buddhism, Hinduism
Languages: Dzongkha, Tibetan dialects, Nepalese dialects
Currency: ngultrum, Indian rupee
Current King: Jigme Khesar Namgyel Wangchuck
Cities (capital, largest, or with at least a million people):
   Thimphu: •located in western Bhutan •capital of Bhutan •most populated city in Bhutan (89,000 people) •located in a valley of the Himalaya

ENVIRONMENTAL/ECONOMIC:
Climate: tropical in the southern plains; cool winters/hot summers in the central valleys; severe winters/cold summers in the high mountains
Natural Resources: timber, hydropower, gypsum, calcium carbide
Agricultural Products: rice, corn, root crops, citrus, dairy products
Major Exports: electricity, cardamom, gypsum, timber, handicrafts
Natural Hazards: violent storms in the mountains, landslides

BOLIVIA
Country Name: Republic of Bolivia
Continent: South America
Area: 424,164 sq mi / 1,098,581 sq km
Population: 10,088,000
Population Density: 24 people per sq mi / 9 people per sq km
Capitals: La Paz (administrative capital), Sucre (judicial capital)

PHYSICAL:
Highest Point: Nevado Sajama 21,463 ft / 6,542 m
Lowest Point: Paraguay River 295 ft / 90 m
Mountain Ranges:
  Andes: •located in western and central Bolivia •extend into Peru, Chile, and Argentina
Peaks (minimum elevation of 8,000 ft / 2,400 m):
  Nevado Sajama: •located in the Andes •highest point in Bolivia

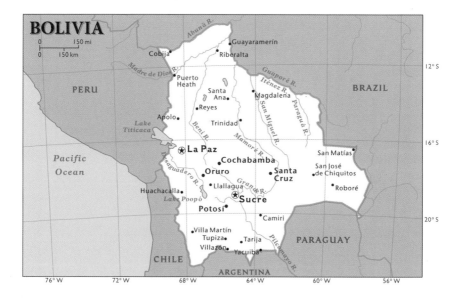

PAGE 52 | THE ULTIMATE FACT BOOK

Plateau Regions:
  Altiplano: •located in southwestern Bolivia •extends into Peru •basin of the Andes
Swamps:
  Pantanal: •located in eastern Bolivia •extends into Brazil and Paraguay •largest swamp in the world
Lakes:
  Poopo: •located in western Bolivia •located on the Altiplano near the slopes of the Andes
  Titicaca: •located in western Bolivia •on the Bolivia-Peru border •on the Altiplano near the slopes of the Andes •largest lake in South America
Rivers:
  Madeira: •forms part of the Bolivia-Brazil border •extends into Brazil •has its source in the Andes
  Mamore: •tributary of the Madeira •forms part of the Bolivia-Brazil border •extends into Brazil •has its source in the Andes

POLITICAL:
Independence: August 6, 1825 (from Spain)
Bordering Countries: Brazil, Peru, Chile, Argentina, Paraguay (5)
Administrative Divisions: Beni, Chuquisaca, Cochabamba, La Paz, Oruro, Pando, Potosi, Santa Cruz, Tarija (9 departments)
Ethnic/Racial Groups: Quechua, mestizo, Aymara, white
Religion: Christianity (primarily Roman Catholic)
Languages: Spanish, Quechua, Aymara
Currency: boliviano
Current President: Juan Evo Morales Ayma
Cities (capital, largest, or with at least a million people):
  La Paz: •located in western Bolivia •city in the La Paz department •administrative capital of Bolivia •most populated city in Bolivia (1,642,000 people)
  Santa Cruz: •located in central Bolivia •city in the Santa Cruz department •on the edge of the Andes
  Sucre: •located in southern Bolivia •city in the Chuquisaca department •judicial capital of Bolivia (281,000 people)

ENVIRONMENTAL/ECONOMIC:
Climate: humid and tropical in the rain forest; cold and semi-arid in the mountains
Natural Resources: tin, natural gas, petroleum, zinc, tungsten
Agricultural Products: soybeans, coffee, coca, cotton, timber
Major Exports: soybeans, natural gas, zinc, gold, wood
Natural Hazards: floods, earthquakes

BOSNIA AND HERZEGOVINA

Country Name: Republic of Bosnia and Herzegovina
Continent: Europe
Area: 19,741 sq mi / 51,129 sq km
Population: 3,843,000
Population Density: 195 people per sq mi / 75 people per sq km
Capital: Sarajevo

PHYSICAL:
Highest Point: Maglic 7,828 ft / 2,386 m
Lowest Point: Adriatic Sea 0 ft / 0 m (sea level)
Mountain Ranges:
    Dinaric Alps: •located in western Bosnia and Herzegovina •extend into Croatia and into Serbia and Montenegro
Rivers:
    Sava: •forms part of Bosnia and Herzegovina's border with Croatia and with Serbia and Montenegro •extends into Croatia and into Serbia and Montenegro

POLITICAL:
Independence: March 1, 1992 (from Yugoslavia)
Former Names: People's Republic of Bosnia and Herzegovina, Socialist Republic of Bosnia and Herzegovina
Bordering Countries: Croatia, Montenegro, and Serbia (3)

Administrative Divisions: Federation of Bosnia and Herzegovina, Republika Srpska (2 first-order administrative divisions) •Brcko (1 internationally supervised district)
Ethnic/Racial Groups: Bosniak, Serbian, Croatian
Religions: Islam, Christianity (Orthodox, Roman Catholic)
Languages: Croatian, Serbian, Bosnian
Currency: marka
Current President (3 presidency members; rotating terms): Zeljko Komsic, Nebojsa Radmanovic, Bakir Izetbegovic
Cities (capital, largest, or with at least a million people):
Sarajevo: •located in central Bosnia and Herzegovina •city in the Bosniak/Croat Federation of Bosnia and Herzegovina first-order administrative division • capital of Bosnia and Herzegovina •most populated city in Bosnia and Herzegovina (392,000 people) •on the edge of the Dinaric Alps

ENVIRONMENTAL/ECONOMIC:
Climate: hot summers/cold winters in most areas; short and cool summers/long and severe winters in the mountains; mild and rainy winters along the coast
Natural Resources: coal, iron ore, bauxite, copper, lead
Agricultural Products: wheat, corn, fruits, vegetables, livestock
Major Exports: metals, clothing, wood products
Natural Hazards: earthquakes

BOTSWANA
Country Name: Republic of Botswana
Continent: Africa
Area: 224,607 sq mi / 581,730 sq km
Population: 2,033,000
Population Density: 9 people per sq mi / 3.5 people per sq km
Capital: Gaborone

PHYSICAL:
Highest Point: point in the Tsodilo Hills 4,885 ft / 1,489 m
Lowest Point: Limpopo/Shashe River Junction 1,683 ft / 513 m
Swamps:
Okavango Delta: •located in northwestern Botswana
Deserts:
Kalahari: •covers most of the country •extends into Namibia and South Africa
Rivers:
Limpopo: •forms part of the Botswana–South Africa border
Okavango: •has its mouth in the Okavango Delta •extends into Namibia
Deltas:
Okavango Delta: •mouth of the Okavango River •feeds into northwestern Botswana

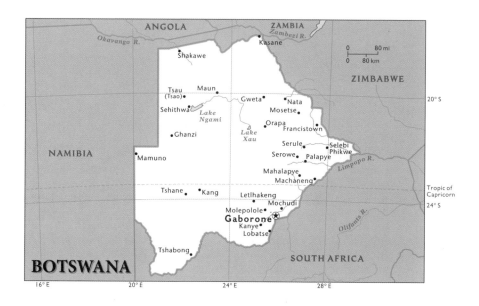

POLITICAL:
Independence: September 30, 1966 (from the United Kingdom)
Former Names: Bechuanaland
Bordering Countries: South Africa, Namibia, Zimbabwe (3)
Administrative Divisions: Central, Ghanzi, Kgalagadi, Kgatleng, Kweneng, Northwest, Northeast, Southeast, Southern (9 districts) •Francistown, Gaborone, Jwaneng, Lobatse, Selebi-Pikwe (5 town councils)
Ethnic/Racial Groups: Tswana, Kalanga, Basarwa
Religions: indigenous beliefs, Christianity
Languages: English, Setswana
Currency: pula
Current President: Seretse Khama Ian Khama
Cities (capital, largest, or with at least a million people):
   Gaborone: •located in southern Botswana •city in the Gaborone town council •capital of Botswana •most populated city in Botswana (196,000 people)

ENVIRONMENTAL/ECONOMIC:
Climate: semi-arid with warm winters/hot summers
Natural Resources: diamonds, copper, nickel, salt, soda ash
Agricultural Products: sorghum, maize, millet, livestock
Major Exports: diamonds, copper, nickel, soda ash, meat
Natural Hazards: droughts, sandstorms, dust storms

BRAZIL

Country Name: Federative Republic of Brazil
Continent: South America
Area: 3,300,169 sq mi / 8,547,403 sq km (largest in South America)
Population: 196,655,000 (largest in South America)
Population Density: 60 people per sq mi / 23 people per sq km
Capital: Brasilia

PHYSICAL:

Highest Point: Pico da Neblina 9,823 ft / 2,994 m
Lowest Point: Atlantic Ocean 0 ft / 0 m (sea level)
Mountain Ranges:
　Brazilian Highlands: ●cover much of eastern Brazil
　Guiana Highlands: ●located in northern Brazil ●extend into Venezuela, Guyana, and Suriname
Peaks (minimum elevation of 8,000 ft / 2,400 m):
　Pico da Neblina: ●on the Brazil-Venezuela border ●highest point in Brazil
Plateau Regions:
　Mato Grosso: ●located in central Brazil ●high region west of the Brazilian Highlands
Rain Forests:
　Amazon: ●located in northern Brazil ●extends into Venezuela, Colombia, Peru, and Bolivia ●largest rain forest in the world

Swamps:
    Pantanal: •located in central Brazil •extends into Bolivia and Paraguay •largest swamp in the world

Oceans:
    Atlantic: •forms the northern and eastern coasts of Brazil

Lagoons:
    Lagoa dos Patos: •located in southeastern Brazil •feeds into the Atlantic Ocean

Rivers:
    Amazon: •has its mouth in the Atlantic Ocean •extends into Peru •longest river in South America •second longest river in the world
    Araguaia: •tributary of the Tocantins •has its source in the Brazilian Highlands
    Iguacu: •tributary of the Parana River •has its source in southern Brazil •forms part of the Brazil-Argentina border
    Japura: •tributary of the Amazon •crosses into Colombia
    Jurua: •tributary of the Amazon •has its source in northwestern Brazil
    Madeira: •tributary of the Amazon •forms part of the Brazil-Bolivia border
    Negro: •tributary of the Amazon
    Paraguai (Paraguay): •has its source in the Mato Grosso Plateau •forms part of the Brazil-Paraguay border •extends into Paraguay
    Parana: •has its source in the Brazilian Highlands •forms part of the Brazil-Paraguay border •extends into Argentina
    Parnaiba: •mouth is in the Atlantic Ocean •source is in the Brazilian Highlands
    Purus: •tributary of the Amazon •extends into Peru
    Sao Francisco: •mouth is in the Atlantic Ocean •source is in the Brazilian Highlands
    Tapajos: •tributary of the Amazon •has its source on the Mato Grosso Plateau
    Teles Pires: •tributary of the Tapajos •has its source on the Mato Grosso Plateau
    Tocantins: •mouth is in the Atlantic Ocean • source is in the Brazilian Highlands
    Uruguai (Uruguay): • source is in southern Brazil •forms part of the Brazil-Argentina border
    Xingu: •tributary of the Amazon •has its source on the Mato Grosso Plateau

Deltas:
    Amazon River Delta: •mouth of the Amazon River •feeds into the Atlantic Ocean •largest river discharge in the world

Waterfalls:
    Iguazu: •located in southwestern Brazil •part of the Iguacu River •on the Brazil-Argentina border

Dams:
    Itaipu: •located in southwestern Brazil •crosses the Parana River •on the Brazil-Paraguay border •forms Itaipu Reservoir

Lakes:
    Itaipu Reservoir: •located in southwestern Brazil •on the Brazil-Paraguay border •fed by the Parana River •formed by the Itaipu Dam

Islands:
  Marajo: •an island in the Amazon River Delta •in northern Brazil •bordered by the Atlantic Ocean and the Amazon River

POLITICAL:
Independence: September 7, 1822 (from Portugal)
Bordering Countries: Bolivia, Venezuela, Colombia, Peru, Paraguay, Argentina, Guyana, Uruguay, French Guiana (overseas territory of France), Suriname (10)
Administrative Divisions: Acre, Alagoas, Amapa, Amazonas, Bahia, Ceara, Espirito Santo, Goias, Maranhao, Mato Grosso, Mato Grosso do Sul, Minas Gerais, Para, Paraiba, Parana, Pernambuco, Piaui, Rio de Janeiro, Rio Grande do Norte, Rio Grande do Sul, Rondonia, Roraima, Santa Catarina, Sao Paulo, Sergipe, Tocantins (26 states) •Distrito Federal (1 federal district)
Ethnic/Racial Groups: European (mainly Portuguese, German, Italian, Spanish), black
Religion: Christianity (primarily Roman Catholic)
Languages: Portuguese, Spanish, English, French
Currency: real
Current President: Dilma Rousseff
Cities (capital, largest, or with at least two million people):
  Sao Paulo: •located in southern Brazil •city in the Sao Paulo state •most populated city in Brazil (20,262,000 people)
  Rio de Janeiro: •located in southeastern Brazil •city in the Rio de Janeiro state •chief port on the Atlantic Ocean
  Salvador: •located in eastern Brazil •city in the Bahia state •a major port on the Atlantic Ocean
  Belo Horizonte: •located in eastern Brazil •city in the Minas Gerais state •located in a valley in the Brazilian Highlands
  Fortaleza: •located in northeastern Brazil •city in the Ceara state •a major port on the Atlantic Ocean
  Brasilia: •located in central Brazil •capital of Brazil (3,789,000 people)•located in the federal district •in the Brazilian Highlands
  Curitiba: •located in southern Brazil •city in the Parana state •chief port on the Iguacu River
  Manaus: •located in northwestern Brazil •city in the Amazonas state •chief port on the Amazon River
  Recife: •located in northeastern Brazil •city in the Pernambuco state •a major port on the Atlantic Ocean
  Belem: •located in northern Brazil •city in the Para state •a major port on the Atlantic Ocean
  Porto Alegre: •located in southwestern Brazil •city in the Rio Grande do Sul state •chief port on the Lagoa dos Patos (lagoon)
  Goiania: •located in central Brazil •city in the Goias state •in the Brazilian Highlands
  Campinas: •located in southern Brazil •city in the Sao Paulo state

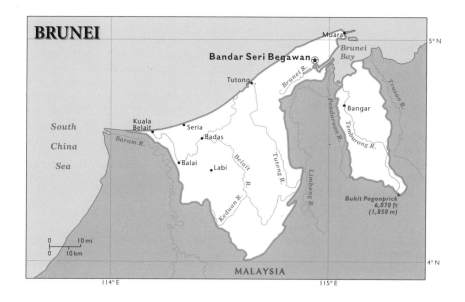

ENVIRONMENTAL/ECONOMIC:
Climate: mostly tropical; temperate in the south
Natural Resources: bauxite, gold, iron ore, manganese, nickel
Agricultural Products: coffee, soybeans, wheat, rice, beef
Major Exports: transport equipment, iron ore, soybeans, footwear, coffee
Natural Hazards: droughts, floods

BRUNEI
Country Name: Negara Brunei Darussalam
Continent: Asia
Area: 2,226 sq mi / 5,765 sq km
Population: 410,000
Population Density: 184 people per sq mi / 71 people per sq km
Capital: Bandar Seri Begawan

PHYSICAL:
Highest Point: Bukit Pagon 6,070 ft / 1,850 m
Lowest Point: South China Sea 0 ft / 0 m (sea level)
Seas:
    South China: •forms the northern coast of Brunei •feeds into the Pacific Ocean
Islands:
    Borneo: •island where Brunei is located •bordered by the South China Sea •also occupied by Malaysia and Indonesia

POLITICAL:
Independence: January 1, 1984 (from the United Kingdom)
Bordering Country: Malaysia (1)
Administrative Divisions: Belait, Brunei and Muara, Temburong, Tutong (4 districts)
Ethnic/Racial Groups: Malay, Chinese, indigenous
Religions: Islam, Buddhism, Christianity, indigenous beliefs
Languages: Malay, English, Chinese
Currency: Bruneian dollar
Current Sultan and Prime Minister: Hassanal Bolkiah
Cities (capital, largest, or with at least a million people):
　Bandar Seri Begawan: • located in northern Brunei • city in the Brunei and Muara district • capital of Brunei • most populated city in Brunei (22,000 people) • chief port on Brunei Bay

ENVIRONMENTAL/ECONOMIC:
Climate: tropical; hot and rainy
Natural Resources: petroleum, natural gas, timber
Agricultural Products: rice, vegetables, fruits, chickens
Major Exports: crude oil, natural gas, refined products
Natural Hazards: typhoons, earthquakes, floods (though rare)

BULGARIA
Country Name: Republic of Bulgaria
Continent: Europe
Area: 42,855 sq mi / 110,994 sq km
Population: 7,476,000
Population Density: 174 people per sq mi / 67 people per sq km
Capital: Sofia

PHYSICAL:
Highest Point: Musala 9,596 ft / 2,925 m
Lowest Point: Black Sea 0 ft / 0 m (sea level)
Mountain Ranges:
　Balkan: • cover central Bulgaria • extend into Serbia and Montenegro
　Pirin: • located in southwestern Bulgaria • extend into Greece
　Rhodope: • located in southern Bulgaria • extend into Greece
　Rila: • located in southwestern Bulgaria
Peaks (minimum elevation of 8,000 ft / 2,400 m):
　Musala: • located in the Rila Mountains • highest point in Bulgaria
Seas:
　Black: • forms the eastern coast of Bulgaria • feeds into the Sea of Marmara

Rivers:
> Danube: •forms most of the Bulgaria-Romania border •extends into Serbia and Montenegro and into Romania

Regions:
> Dobruja: •lowland region in northeastern Bulgaria •extends into Romania •located west of the Black Sea
> Macedonia: •mountainous region in southwestern Bulgaria •extends into Macedonia and Greece
> Thrace: •lowland region in southern Bulgaria •extends into Greece and Turkey

POLITICAL:
Independence: March 3, 1878 (from the Ottoman Empire)
Bordering Countries: Romania, Greece, Serbia, Turkey, Macedonia (5)
Administrative Divisions: Blagoevgrad, Burgas, Dobrich, Gabrovo, Khaskovo, Kurdzhali, Kyustendil, Lovech, Montana, Pazardzhik, Pernik, Pleven, Plovdiv, Razgrad, Ruse, Shumen, Silistra, Sliven, Smolyan, Sofiya, Sofiya-Grad, Stara Zagora, Turgovishte, Varna, Veliko Turnovo, Vidin, Vratsa, Yambol (28 provinces)
Ethnic/Racial Groups: Bulgarian, Turkish
Religions: Christianity (primarily Bulgarian Orthodox), Islam
Language: Bulgarian
Currency: lev
Current President: Rosen Plevnehev

Cities (capital, largest, or with at least a million people):
Sofia: •located in western Bulgaria •city in the Sofia province •capital of Bulgaria •most populated city in Bulgaria (1,192,000 people) •on the edge of the Balkan and the Rhodope Mountains

ENVIRONMENTAL/ECONOMIC:
Climate: temperate; cold and damp winters/hot and dry summers
Natural Resources: bauxite, copper, lead, zinc, coal
Agricultural Products: vegetables, fruits, tobacco, livestock
Major Exports: clothing, footwear, iron and steel, machinery and equipment
Natural Hazards: earthquakes, landslides

BURKINA FASO
Country Name: Burkina Faso
Continent: Africa
Area: 105,869 sq mi / 274,200 sq km
Population: 16,968,000
Population Density: 160 people per sq mi / 62 people per sq km
Capital: Ouagadougou

PHYSICAL:
Highest Point: Tena Kourou 2,457 ft / 749 m
Lowest Point: Black Volta River 656 ft / 200 m

Grasslands/Prairies:
  Sahel: •covers most of country •extends into Mali, Ghana, Togo, Benin, and Niger
Rivers:
  Black Volta: •has its source in southwestern Burkina Faso •forms part of the Burkina Faso-Ghana border •extends into Ghana
  White Volta: •has its source in the Sahel of northern Burkina Faso •extends into Ghana

POLITICAL:
Independence: August 5, 1960 (from France)
Former Name: Upper Volta
Bordering Countries: Mali, Niger, Cote d'Ivoire, Ghana, Benin, Togo (6)
Administrative Divisions: Bale, Bam, Banwa, Bazega, Bougouriba, Boulgou, Boulkiemde, Comoe, Ganzourgou, Gnagna, Gourma, Houet, Ioba, Kadiogo, Kenedougou, Komondjari, Kompienga, Kossi, Koulpelogo, Kouritenga, Kourweogo, Leraba, Loroum, Mouhoun, Nahouri, Namentenga, Nayala, Noumbiel, Oubritenga, Oudalan, Passore, Poni, Sanguie, Sanmatenga, Seno, Sissili, Soum, Sourou, Tapoa, Tuy, Yagha, Yatenga, Ziro, Zondoma, Zoundweogo (45 provinces)
Ethnic/Racial Groups: Mossi, Gurunsi, Senufo, Lobi, Bobo, Mande, Fulani
Religions: Islam, indigenous beliefs, Christianity (mostly Roman Catholic)
Languages: French, indigenous African languages
Currency: Communaute Financiere Africaine franc
Current President: Blaise Compaore
Cities (capital, largest, or with at least a million people):
  Ouagadougou: •located in central Burkina Faso •city in the Kadiogo province •capital of Burkina Faso •most populated city in Burkina Faso (1,777,000 people)

ENVIRONMENTAL/ECONOMIC:
Climate: tropical; warm and dry winters/hot and wet summers
Natural Resources: manganese, limestone, marble, gold, antimony
Agricultural Products: cotton, peanuts, shea nuts, sesame, livestock
Major Exports: cotton, livestock, gold
Natural Hazards: droughts

BURUNDI
Country Name: Republic of Burundi
Continent: Africa
Area: 10,747 sq mi / 27,834 sq km
Population: 10,216,000
Population Density: 951 people per sq mi / 367 people per sq km
Capital: Bujumbura

PHYSICAL:
Highest Point: Mount Heha 8,760 ft / 2,670 m
Lowest Point: Lake Tanganyika 2,533 ft / 772 m
Mountain Ranges:
    Mitumba: •cover western Burundi •extend into Rwanda, Democratic Republic of the Congo, and Tanzania
Peaks (minimum elevation of 8,000 ft / 2,400 m):
    Mount Heha: •located in the Mitumba Mountains •highest point in Burundi
Lakes:
    Tanganyika: •located in southwestern Burundi •on the border with Democratic Republic of the Congo and Tanzania

POLITICAL:
Independence: July 1, 1962 (from Belgian administration)
Former Name: Urundi
Bordering Countries: Tanzania, Rwanda, Democratic Republic of the Congo (3)
Administrative Divisions: Bubanza, Bujumbura Mairie, Bujumbura Rural, Bururi, Cankuzo, Cibitoke, Gitega, Karuzi, Kayanza, Kirundo, Makamba, Muramvya, Muyinga, Mwaro, Ngozi, Rutana, Ruyigi (17 provinces)
Ethnic/Racial Groups: Hutu, Tutsi
Religions: Christianity (Roman Catholic, Protestant), indigenous beliefs, Islam
Languages: Kirundi, French, Swahili
Currency: Burundi franc
Current President: Pierre Nkurunziza

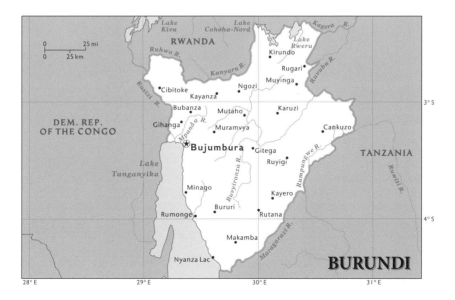

Cities (capital, largest, or with at least a million people):
  Bujumbura: •located in western Burundi •city in the Bujumbura province •capital of Burundi •most populated city in Burundi (455,000 people) •chief port on Lake Tanganyika

ENVIRONMENTAL/ECONOMIC:
Climate: equatorial with wet season and dry season
Natural Resources: nickel, uranium, rare earth oxides, peat, cobalt
Agricultural Products: coffee, cotton, tea, corn, beef
Major Exports: coffee, tea, sugar, cotton, hides
Natural Hazards: flooding, landslides, droughts

COUNTRIES

CAMBODIA
Country Name: Kingdom of Cambodia
Continent: Asia
Area: 69,898 sq mi / 181,035 sq km
Population: 14,702,000
Population Density: 210 people per sq mi / 81 people per sq km
Capital: Phnom Penh

PHYSICAL:
Highest Point: Phnom Aural, 5,948 ft / 1,813 m
Lowest Point: Gulf of Thailand 0 ft / 0 m (sea level)
Gulfs:
  Gulf of Thailand: •forms the southwestern coast of Cambodia •feeds into the South China Sea
Lakes:
  Tonle Sap: •located in western Cambodia
Rivers:
  Mekong: •extends into Laos and Vietnam

COUNTRIES A TO Z | PAGE 67

POLITICAL:
Independence: November 9, 1953 (from France)
Former Names: Khmer Republic, Democratic Kampuchea, People's Republic of Kampuchea, State of Cambodia
Bordering Countries: Vietnam, Thailand, Laos (3)
Administrative Divisions: Banteay Mean Choay, Batdambang, Kampong Cham, Kampong Chhnang, Kampong Spoe, Kampong Thom, Kampot, Kandal, Kaoh Kong, Keb, Kracheh, Mondol Kiri, Otdar Mean Cheay, Pailin, Pouthisat, Preah Seihanu, Preah Vihear, Prey Veng, Rotanokiri, Siem Reab, Stoeng Treng, Svay Rieng, Takev (23 provinces) • Phnom Penh (1 municipality)
Ethnic/Racial Groups: Khmer, Vietnamese
Religion: Theravada Buddhism
Languages: Khmer, French, English
Currency: riel
Current Prime Minister: Hun Sen
Cities (capital, largest, or with at least a million people):
  Phnom Penh: • located in southern Cambodia • city in the Phnom Penh municipality • capital of Cambodia • most populated city in Cambodia (1,519,000 people) • chief port on the Mekong River

ENVIRONMENTAL/ECONOMIC:
Climate: tropical with a rainy monsoon season and a dry season
Natural Resources: oil and gas, timber, gemstones, iron ore, manganese
Agricultural Products: rice, rubber, corn, vegetables
Major Exports: timber, garments, rubber, rice, fish
Natural Hazards: monsoons, floods, droughts

CAMEROON
Country Name: Republic of Cameroon
Continent: Africa
Area: 183,569 sq mi / 475,442 sq km
Population: 20,052,000
Population Density: 109 people per sq mi / 42 people per sq km
Capital: Yaounde

PHYSICAL:
Highest Point: Fako (on Cameroon Mountain) 13,435 ft / 4,095 m
Lowest Point: Gulf of Guinea 0 ft / 0 m (sea level)
Mountain Ranges:
  Adamawa: • covers much of central Cameroon • extends into Nigeria
Peaks (minimum elevation of 8,000 ft / 2,400 m):
  Cameroon Mountain: • located in southwestern Cameroon • includes Fako, the highest point in Cameroon

Grasslands/Prairies:
  Sahel: •located in northern Cameroon •extends into Nigeria, Chad, and Central African Republic
Bays:
  Bight of Bonny: •forms the southwestern coast of Cameroon •feeds into the Gulf of Guinea
Lakes:
  Chad: •located in northern Cameroon •on the Cameroon-Chad border •located in the Sahel
Rivers:
  Benoue (Benue): •has its source in the Adamawa •extends into Nigeria
  Chari: •has its mouth in Lake Chad •forms part of the Cameroon-Chad border •extends into Chad

POLITICAL:

Independence: January 1, 1960 (for French-ruled areas); October 1, 1961 (for British-ruled areas)
Former Names: French Cameroon, British Cameroon
Bordering Countries: Nigeria, Chad, Central African Republic, Republic of the Congo, Gabon, Equatorial Guinea (6)
Administrative Divisions: Adamaoua, Centre, Est, Extreme-Nord, Littoral, Nord, Nord-Ouest, Ouest, Sud, Sud-Ouest (10 provinces)
Ethnic/Racial Groups: Cameroon Highlanders, Equatorial Bantu, Kirdi, Fulani, Northwestern Bantu, Eastern Nigritic, other African groups

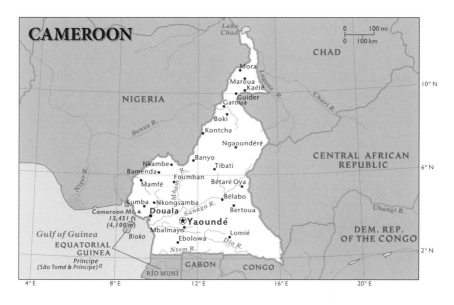

Religions: indigenous beliefs, Christianity, Islam
Languages: French, English, 24 major African language groups
Currency: Communaute Financiere Africaine franc
Current President: Paul Biya
Cities (capital, largest, or with at least a million people):
    Yaounde: •located in southern Cameroon •city in the Centre province •capital of Cameroon (1,739,000 people)
    Douala: •located in southwestern Cameroon •city in the Littoral province •chief port on the Bight of Bonny (2,125,000 people)

ENVIRONMENTAL/ECONOMIC:
Climate: tropical along coast; semi-arid and hot in the north
Natural Resources: petroleum, bauxite, iron ore, timber, hydropower
Agricultural Products: coffee, cacao, cotton, rubber, livestock
Major Exports: crude oil and petroleum products, lumber, cacao, aluminum
Natural Hazards: volcanoes, volcanic gases

CANADA
Country Name: Canada
Continent: North America
Area: 3,855,101 sq mi / 9,984,670 sq km (largest in North America)
Population: 34,468,000
Population Density: 9 people per sq mi / 3.5 people per sq km (least dense in North America)
Capital: Ottawa

PHYSICAL:
Highest Point: Mount Logan 19,551 ft / 5,959 m
Lowest Point: Atlantic Ocean 0 ft / 0 m (sea level)
Mountain Ranges:
    Cassiar: •located in western Canada
    Coast: •cover western Canada •extend into the United States
    Columbia: •located in southwestern Canada
    Laurentian: •located in southeastern Canada
    Mackenzie: •located in northwestern Canada
    Notre Dame: •located in southeastern Canada
    Rocky: •cover western Canada •extend into the United States
    Saint Elias: •located in western Canada •extend into the United States
    Selwyn: •located in northwestern Canada
Peaks (minimum elevation of 8,000 ft / 2,400 m):
    Mount Logan: •located in the Saint Elias Mountains •highest point in Canada
    Mount Robson: •located in the Rocky Mountains

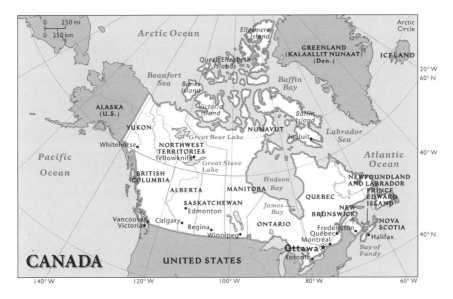

Passes:
    Chilkoot: •located in the Coast Mountains •crosses the border into the United States (Alaska)

Grasslands/Prairies:
    Great Plains: •located in southern Canada •extend into the United States

Oceans:
    Arctic: •forms part of the northern coast of Canada
    Atlantic: •forms part of the eastern coast of Canada
    Pacific: •forms part of the western coast of Canada

Seas:
    Beaufort: •forms part of the northwestern coast of Canada •feeds into the Arctic Ocean
    Labrador: •forms part of the eastern coast of Canada •feeds into the Atlantic Ocean

Gulfs:
    Amundsen: •forms part of the northwestern coast of Canada •feeds into the Beaufort Sea
    Gulf of Boothia: •forms part of the northern coast of Canada
    Gulf of Maine: •forms part of the southeastern coast of Canada •feeds into the Atlantic Ocean
    Gulf of Saint Lawrence: •forms part of the southeastern coast of Canada •feeds into the Atlantic Ocean
    Queen Maud: •forms part of the northern coast of Canada

Bays:
    Baffin: •forms part of the northern coast of Canada •feeds into the Labrador Sea
    Bay of Fundy: •forms part of the southeastern coast of Canada •feeds into the Gulf of Maine •has world's greatest tides

Chaleur: •forms part of the southeastern coast of Canada •feeds into the Gulf of Saint Lawrence
Foxe Basin: •forms part of the northern border of Canada •feeds into Hudson Bay
Georgian: •feeds into Lake Huron
Hudson: •forms part of the central coast of Canada •feeds into the Labrador Sea
James: •forms part of the central coast of Canada •feeds into Hudson Bay
Ungava: •forms part of the eastern coast of Canada •feeds into the Labrador Sea

Sounds:
Queen Charlotte: •forms part of the southwestern coast of Canada •feeds into the Pacific Ocean

Straits:
Cabot: •forms part of the southeastern coast of Canada •separates the Island of Newfoundland from Cape Breton Island •connects the Gulf of Saint Lawrence and the Atlantic Ocean
Davis: •forms part of the northeastern coast of Canada •separates Canada from Greenland (territory of Denmark) •connects Baffin Bay and the Labrador Sea
Hudson: •forms part of the eastern coast of Canada •separates Baffin Island from mainland Canada •connects Hudson Bay and the Labrador Sea
Strait of Georgia: •forms part of the southwestern coast of Canada •separates Vancouver Island from mainland Canada •connects Queen Charlotte Sound and Puget Sound
Strait of Juan de Fuca: •forms part of the southwestern coast of Canada •separates Vancouver Island from the United States •connects Puget Sound and the Pacific Ocean

Channels:
Foxe: •forms part of the northern coast of Canada •separates Baffin Island from Southampton Island •connects Foxe Basin and Hudson Bay
Parry: •forms part of the northern coast of Canada •separates Melville and Devon Islands from Victoria, Prince of Wales, and Baffin Islands •connects the Arctic Ocean and Baffin Bay

Canals:
Welland: •connects Lake Erie and Lake Ontario

Lakes:
Great Bear: •located in northwestern Canada •largest lake entirely in Canada •eighth largest lake in the world
Great Slave: •located in western Canada •fed by the Slave River •tenth largest lake in the world
Lake Athabasca: •located in central Canada
Lake Erie: •located in southern Canada •on the Canada–United States border
Lake Huron: •located in southern Canada •on the Canada–United States border •fourth largest lake in the world •third largest of the Great Lakes in surface area
Lake Manitoba: •located in southern Canada
Lake Nipigon: •located in southern Canada
Lake Ontario: •located in southern Canada •on the Canada–United States border

Lake Superior: •located in southern Canada •on the Canada–United States border •westernmost of the Great Lakes •largest lake in North America •second largest lake in the world
Lake Winnipeg: •located in southern Canada
Lake Winnipegosis: •located in southern Canada
Reindeer: •located in central Canada

Rivers:
Columbia: •has its source in the Rocky Mountains •extends into the United States
Fraser: •has its mouth in the Strait of Georgia •has its source in the Rocky Mountains
Mackenzie: •has its mouth in the Beaufort Sea •has its source in the Rocky Mountains
Nelson: •has its mouth in Hudson Bay •has its source in the Rocky Mountains
North Saskatchewan: •tributary of the Saskatchewan River •has its source in the Rocky Mountains
Ottawa: •tributary of the Saint Lawrence River •has its source in southern Canada
Peace: •tributary of the Slave River •has its source in the Rocky Mountains
Saint Lawrence: •has its mouth in the Gulf of Saint Lawrence •forms part of the Canada–United States border
Saskatchewan: •tributary of the Nelson River •has its source in the Rocky Mountains
Slave: •tributary of the Mackenzie River
South Saskatchewan: •tributary of the Saskatchewan River •has its source in the Rocky Mountains
Yukon: •has its source in the Coast Mountains •extends into the United States (Alaska)

Deltas:
Mackenzie River Delta: •mouth of the Mackenzie River •feeds into the Beaufort Sea

Waterfalls:
Niagara: •located in southeastern Canada •part of the Niagara River •on the Canada–United States border (between Ontario and New York)

Peninsulas:
Gaspe: •bordered by the Saint Lawrence River, Chaleur Bay, and Gulf of Saint Lawrence
Nova Scotia: •bordered by the Atlantic Ocean, Gulf of Maine, and Bay of Fundy
Ungava: •bordered by Hudson Bay, Hudson Strait, and Ungava Bay

Islands:
Baffin: •an island off the northern coast of mainland Canada •bordered by Baffin Bay, Parry Channel, Gulf of Boothia, Foxe Basin, Foxe Channel, Hudson Strait, and Davis Strait •fifth largest island in the world
Banks: •an island off the northwestern coast of mainland Canada •bordered by the Arctic Ocean, Beaufort Sea, and Amundsen Gulf
Cape Breton: •an island off the southeastern coast of mainland Canada •bordered by the Gulf of Saint Lawrence, Atlantic Ocean, and Cabot Strait
Devon: •an island in the Queen Elizabeth Islands in northern Canada bordered by Parry Channel and Baffin Bay
Ellesmere: •an island in the Queen Elizabeth Islands in northern Canada bordered by the Arctic Ocean and Baffin Bay •tenth largest island in the world

Melville: •an island in the Queen Elizabeth Islands in northern Canada •bordered by the Parry Channel
Newfoundland: •an island off the eastern coast of mainland Canada •bordered by the Atlantic Ocean, Gulf of Saint Lawrence, and Cabot Strait
Prince Edward: •an island off the southeastern coast of mainland Canada •bordered by the Gulf of Saint Lawrence •also a province
Prince of Wales: •an island off the northern coast of mainland Canada •bordered by Parry Channel
Queen Charlotte: •an island group west of mainland Canada •bordered by the Pacific Ocean and Queen Charlotte Sound
Southampton: •an island off the northern coast of mainland Canada •bordered by the Foxe Channel, Foxe Basin, and Hudson Bay
Vancouver: •an island off the southwestern coast of mainland Canada •bordered by the Pacific Ocean, Queen Charlotte Sound, Strait of Georgia, and Strait of Juan de Fuca
Victoria: •an island off the northern coast of mainland Canada •bordered by the Amundsen Gulf, Parry Channel, and Queen Maud Gulf •ninth largest island in the world

POLITICAL:
Independence: July 1, 1867 (from the United Kingdom); complete legal autonomy 1931
Bordering Country: United States (1)
Administrative Divisions: Alberta, British Columbia, Manitoba, New Brunswick, Newfoundland and Labrador, Nova Scotia, Ontario, Prince Edward Island, Quebec, Saskatchewan (10 provinces) •Northwest Territories, Nunavut, Yukon (3 territories)
Ethnic/Racial Groups: British, French, other European groups, Amerindian
Religion: Christianity (Roman Catholic, Protestant)
Languages: English, French
Currency: Canadian dollar
Current Prime Minister: Stephen Harper
Cities (capital, largest, or with at least a million people):
Toronto: •located in southern Canada •city in the Ontario province •most populated city in Canada (5,449,000 people) •chief port on Lake Ontario
Montreal: •located in southeastern Canada •city in the Quebec province •chief port on the Saint Lawrence River
Vancouver: •located in southwestern Canada •city in the British Columbia province •chief port on the Strait of Georgia
Calgary: •located in southwestern Canada •city in the Alberta province
Ottawa: •located in southeastern Canada •city in the Ontario province •capital of Canada (1,170,000 people) •chief port on the Ottawa River
Edmonton: •located in southwestern Canada •city in the Alberta province

ENVIRONMENTAL/ECONOMIC:
Climate: temperate in south, subarctic to arctic in the north

Natural Resources: iron ore, nickel, zinc, copper, gold
Agricultural Products: wheat, barley, oilseed, tobacco, dairy products
Major Exports: motor vehicles and parts, industrial machinery, aircraft, telecommunications equipment, chemicals
Natural Hazards: thunderstorms

CAPE VERDE

Country Name: Republic of Cape Verde
Continent: Africa
Area: 1,558 sq mi / 4,036 sq km
Population: 496,000
Population Density: 318 people per sq mi / 123 people per sq km
Capital: Praia

PHYSICAL:

Highest Point: Mount Fogo 9,281 ft / 2,829 m
Lowest Point: Atlantic Ocean 0 ft / 0 m (sea level)
Peaks (minimum elevation of 8,000 ft / 2,400 m):
  Mount Fogo: •located on the island of Fogo •highest point in Cape Verde
Oceans:
  Atlantic: •forms the entire coast of Cape Verde
Islands:
  Santiago: •largest island in Cape Verde •surrounded by the Atlantic Ocean
  Santo Antao: •second largest island in Cape Verde •surrounded by the Atlantic Ocean

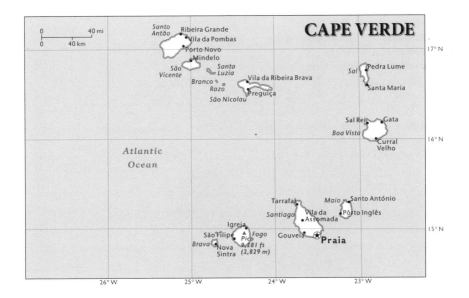

POLITICAL:
Independence: July 5, 1975 (from Portugal)
Administrative Divisions: Boa Vista, Brava, Maio, Mosteiros, Paul, Porto Novo, Praia, Ribeira Brava, Ribeira Grande, Ribeira Grande de Santiago, Sal, Santa Catarina, Santa Catarina do Fogo, Santa Cruz, Sao Domingos, Sao Filipe, Sao Lourenco dos Orgaos, Sao Miguel, Sao Salvador do Mundo, Sao Vicente, Tarrafal, Tarrafal de Sao Nicolau (22 municipalities)
Ethnic/Racial Groups: Creole, African
Religion: Christianity (Roman Catholic, Protestant)
Languages: Portuguese, Crioulo (Kriolu)
Currency: Cape Verdean escudo
Current President: Jorge Carlos Fonseca
Cities (capital, largest, or with at least a million people):
  Praia: •located on Santiago Island •city in the Praia municipality •capital of Cape Verde •most populated city in Cape Verde (125,000 people) •chief port on the Atlantic Ocean

ENVIRONMENTAL/ECONOMIC:
Climate: temperate, warm and dry summers
Natural Resources: salt, basalt rock, limestone, kaolin, fish
Agricultural Products: bananas, corn, beans, sweet potatoes
Natural Disasters: droughts, harmattan winds, dust storms, volcanoes, earthquakes

CENTRAL AFRICAN REPUBLIC
Country Name: Central African Republic
Continent: Africa
Area: 240,535 sq mi / 622,984 sq km
Population: 4,950,000
Population Density: 21 people per sq mi / 8 people per sq km
Capital: Bangui

PHYSICAL:
Highest Point: Mont Ngaoui 4,659 ft / 1,420 m
Lowest Point: Oubangui River 1,099 ft / 335 m
Grasslands/Prairies:
  Sahel: •located in northern Central African Republic •extends into Cameroon, Chad, and Sudan
Rivers:
  Ubangi (Oubangui): •forms part of the Central African Republic-Democratic Republic of the Congo border

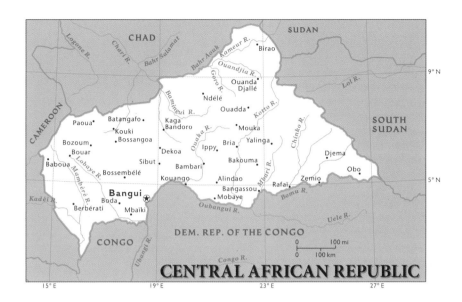

POLITICAL:
Independence: August 16, 1960 (from France)
Former Names: Ubangi-Shari, Central African Empire
Bordering Countries: Democratic Republic of the Congo, Chad, South Sudan, Sudan, Cameroon, Republic of the Congo (6)
Regions:
  Dar Rounga: • located in northern Central African Republic • extends into Sudan
Administrative Divisions: Bamingui-Bangoran, Basse-Kotto, Haute-Kotto, Haut-Mbomou, Kemo, Lobaye, Mambere-Kadei, Mbomou, Nana-Mambere, Ombella-Mpoko, Ouaka, Ouham, Ouham-Pende, Vakaga (14 prefectures) • Nana-Grebizi, Sangha-Mbaere (2 economic prefectures) • Bangui (1 commune)
Ethnic/Racial Groups: Baya, Banda, Mandjia, Sara, Mboum, M'Baka, Yakoma
Religions: indigenous beliefs, Christianity (Protestant, Roman Catholic), Islam
Languages: French, Sangho, Arabic, tribal languages
Currency: Communaute Financiere Africaine franc
Current President: Francois Bozize
Cities (capital, largest, or with at least a million people):
  Bangui: • located in southwestern Central African Republic • city in the Bangui commune • capital of Central African Republic • most populated city in Central African Republic (702,000 people) • chief port on the Ubangi River

ENVIRONMENTAL/ECONOMIC:
Climate: tropical with hot and dry winters/mild to hot and wet summers
Natural Resources: diamonds, uranium, timber, gold, oil
Agricultural Products: cotton, coffee, tobacco, manioc (tapioca), timber

Major Exports: diamonds, timber, cotton, coffee, tobacco
Natural Hazards: harmattan winds, floods

CHAD
Country Name: Republic of Chad
Continent: Africa
Area: 495,755 sq mi / 1,284,000 sq km
Population: 11,536,000
Population Density: 23 people per sq mi / 9 people per sq km
Capital: N'Djamena

PHYSICAL:
Highest Point: Emi Koussi 11,204 ft / 3,415 m
Lowest Point: Djourab Depression 525 ft / 160 m
Mountain Ranges:
   Tibesti: •located in northwestern Chad •extend into Libya
Peaks (minimum elevation of 8,000 ft / 2,400 m):
   Emi Koussi: •located in the Tibesti Mountains •highest point in Chad
Grasslands/Prairies:
   Sahel: •covers southern and central Chad •extends into Niger, Nigeria, Cameroon, Central African Republic, and Sudan
Deserts:
   Sahara: •stretches across northern Chad •extends into neighboring Niger, Libya, and Sudan •largest desert in the world

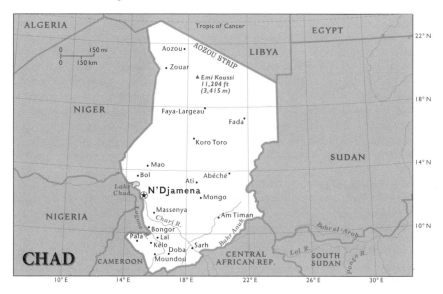

Lakes:
  Lake Chad: •located in western Chad •on the Chad-Cameroon border •located in the Sahel
Rivers:
  Chari: •has its mouth in Lake Chad •forms part of the Chad-Cameroon border
Regions:
  Borkou: •desert region in northern Chad •located to the south of the Tibesti Mountains in the Sahara

POLITICAL:

Independence: August 11, 1960 (from France)
Bordering Countries: Sudan, Central African Republic, Niger, Cameroon, Libya, Nigeria (6)
Regions:
  Aozou Strip: •covers much of northern Chad
Administrative Divisions: Barh el Gazel, Batha, Borkou, Chari-Baguirmi, Ennedi, Guera, Hadjer-Lamis, Kanem, Lac, Logone Occidental, Logone Oriental, Mandoul, Mayo-Kebbi Est, Mayo-Kebbi Ouest, Moyen-Chari, Ouaddai, Salamat, Sila, Tandjile, Tibesti, Ville de N'Djamena, Wadi Fira (22 regions)
Ethnic/Racial Groups: some 200 groups (largest are Arabs and Sara)
Religions: Islam, Christianity, animism
Languages: French, Arabic, Sara, more than 120 different dialects
Currency: Communaute Financiere Africaine franc
Current President: Idriss Deby Itno
Cities (capital, largest, or with at least a million people):
  N'Djamena: •located in southwestern Chad •city in the Chari-Baguirmi prefecture •capital of Chad •most populated city in Chad (808,000 people) •chief port on the Chari River

ENVIRONMENTAL/ECONOMIC:

Climate: tropical in the south; desert in the north
Natural Resources: oil, uranium, natron, kaolin, fish
Agricultural Products: cotton, sorghum, millet, peanuts, cattle
Major Exports: cotton, cattle, gum arabic
Natural Hazards: harmattan winds, droughts, locust plagues

CHILE

Country Name: Republic of Chile
Continent: South America
Area: 291,930 sq mi / 756,096 sq km
Population: 17,268,000
Population Density: 59 people per sq mi / 23 people per sq km
Capital: Santiago

PHYSICAL:

Highest Point: Nevado Ojos del Salado 22,572 ft / 6,880 m
Lowest Point: Pacific Ocean 0 ft / 0 m (sea level)
Mountain Ranges:
   Andes: •cover eastern Chile north to south •extend into Argentina
Peaks (minimum elevation of 8,000 ft / 2,400 m):
   Nevado Ojos del Salado: •located in the Andes •on the Chile-Argentina border •highest point in Chile
Desert:
   Atacama: •located in northwestern Chile •extends into Peru •driest place in the world
Oceans:
   Pacific: •forms most of the western coast of Chile
Gulfs:
   Gulf of Corcovado: •forms part of the central coast of Chile •feeds into the Pacific Ocean
   Gulf of Penas: •forms part of the western coast of Chile •feeds into the Pacific Ocean
Straits:
   Strait of Magellan: •forms part of the southern coast of Chile •separates the island of Tierra del Fuego from mainland Chile •connects the Pacific and Atlantic Oceans
Islands:
   Tierra del Fuego: •group of islands south of mainland Chile •also largest island in the group •bordered by the Pacific Ocean •eastern half owned by Argentina

POLITICAL:
Independence: September 18, 1810 (from Spain)
Bordering Countries: Argentina, Bolivia, Peru (3)
Administrative Divisions: Aisen del General Carlos Ibanez del Campo, Antofagasta, Araucania, Arica y Parinacota, Atacama, Biobio, Coquimbo, Libertador General Bernardo O'Higgins, Los Lagos, Los Rios, Magallanes y de la Antartica Chilena, Maule, Region Metropolitana (Santiago), Tarapaca, Valparaiso (15 regions)
Ethnic/Racial Groups: European, mestizo, Amerindian
Religion: Christianity (Roman Catholic, Protestant)
Language: Spanish
Currency: Chilean peso
Current President: Sebastian Pinera Echenique
Cities (capital, largest, or with at least a million people):
 Santiago: •located in central Chile •city in the Region Metropolitana region •capital of Chile •most populated city in Chile (5,883,000 people) •on the edge of the Andes

ENVIRONMENTAL/ECONOMIC:
Climate: temperate with desert in the north, Mediterranean climate in the middle, cool and damp in the south
Natural Resources: copper, timber, iron ore, nitrates, precious metals (gold, silver)
Agricultural Products: wheat, corn, grapes, beans, beef
Major Exports: copper, fish, fruits, paper and pulp, chemicals
Natural Hazards: earthquakes, volcanoes, tsunamis

CHINA
Country Name: People's Republic of China
Continent: Asia
Area: 3,705,405 sq mi / 9,596,960 sq km
Population: 1,345,855,000 (largest in the world)
Population Density: 363 people per sq mi / 140 people per sq km
Capital: Beijing

PHYSICAL:
Highest Point: Mount Everest 29,035 ft / 8,850 m
Lowest Point: Turpan Depression 505 ft / 154 m below sea level
Mountain Ranges:
 Altay: •located in northwestern China •extend into Kazakhstan, Russia, and Mongolia
 Altun Shan: •located in western China
 Bayan Har Shan: •located in central China
 Gangdise Shan: •located in western China
 Hengduan Shan: •located in southern China
 Himalaya: •cover western China •extend into Pakistan, India, Nepal, and Bhutan

Hindu Kush: •located in western China •extend into Afghanistan, Tajikistan, and Pakistan
Karakoram: •located in western China •extends into Pakistan and India
Kunlun Shan: •located in western China
Qilian Shan: •located in central China
Tian Shan: •located in northwestern China •extend into Kazakhstan and Kyrgyzstan

Peaks (minimum elevation of 8,000 ft / 2,400 m):
K2: •located in the Karakoram Range •on the China-Pakistan border
Mount Everest: •located in the Himalaya •on the China-Nepal border •highest point in the world
Pobedy Peak: •located in the Tian Shan •on the China-Kyrgyzstan border
Xixabangma Feng: •located in the Himalaya

Passes:
Karakoram: •located in the Karakoram Range •crosses the border into India

Plateau Regions:
Dzungarian Basin: •located in northwestern China •basin between the Altay and Tian Shan
Nei Mongol Plateau: •covers northern China •extends into Mongolia •desert plateau north of the Huang He (Yellow River)
Plateau of Tibet: •located in western China •high plateau between the Himalaya and Kunlun Shan •largest plateau in the world
Tarim Basin: •located in western China •basin between the Tian Shan, Kunlun Shan, and Altun Shan

Depressions:
Turpan Depression: •located in northwestern China •lowest point in China

Deserts:
Gobi: •located in northern China •extends into Mongolia •covers part of the Nei Mongol Plateau
Taklimakan: •located in western China •covers most of the Tarim Basin

Seas:
East China: •forms part of the eastern coast of China •feeds into the Philippine Sea
South China: •forms part of the southeastern coast of China •feeds into the Philippine Sea •second largest sea in the world
Yellow: •forms part of the eastern coast of China •feeds into the East China Sea

Gulfs:
Bo Hai: •forms part of the eastern coast of China •feeds into Korea Bay
Gulf of Tonkin: •forms part of the southern coast of China •feeds into the South China Sea

Bays:
Korea: •forms part of the northwestern coast of China •feeds into the Yellow Sea

Straits:
Hainan: •forms part of the southern coast of China •separates the island of Hainan from mainland China •connects the Gulf of Tonkin and the South China Sea
Taiwan: •forms part of the southeastern coast of China •separates the island of Taiwan

from mainland China •connects the East China and South China Seas

Canals:
Grand: •connects the Huang (Yellow River) and Chang Jiang (Yangtze River)

Lakes:
Lop Nur: •located in western China •at the eastern end of the Tarim Basin
Qinghai Hu: •located in central China •near the Qilian Shan

Rivers:
Amur: •forms part of the China-Russia border •extends into Russia •eighth longest river in the world
Brahmaputra: •has its source in the Gangdise Shan •extends into India
Chang Jiang (Yangtze): •has its mouth in the East China Sea •has its source in the Kunlun Shan •third longest river in the world •longest river in Asia
Huang (Yellow): •has its mouth in Bo Hai •has its source in the Bayan Har Shan •sixth longest river in the world
Ertix (Irtysh): •located in northwestern China •extends into Kazakhstan and Russia
Mekong: •has its source on the Plateau of Tibet •extends into Laos and Vietnam
Red: •has its source in the Hengduan Shan •forms part of the China-Vietnam border •extends into Vietnam
Salween: •has its source on the Plateau of Tibet •extends into Myanmar
Xi: •has its mouth in the South China Sea •has its source in southern China

Deltas:
Mouth of the Chang Jiang (Yangtze River): •feeds into the East China Sea

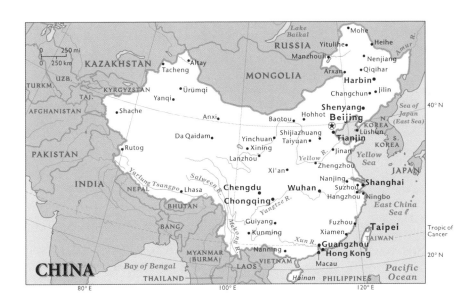

Dams:
   Three Gorges Dam: •located in central China •spans the Chang Jiang (Yangtze River) •world's largest hydroelectric dam
Peninsulas:
   Shandong: •bordered by Bo Hai, Korea Bay, and the Yellow Sea
Islands:
   Hainan: •island off the southern coast of mainland China •bordered by the Gulf of Tonkin, Hainan Strait, and the South China Sea
Regions:
   Manchuria: •plains region in northeastern China •located to the west of the Nei Mongol Plateau

POLITICAL:
Independence: 221 B.C. (unification under the Qin or Chin Dynasty)
Bordering Countries: Mongolia, Russia, India, Myanmar, Kazakhstan, North Korea, Vietnam, Nepal, Kyrgyzstan, Pakistan, Bhutan, Laos, Tajikistan, Afghanistan (14)
Administrative Divisions: Anhui, Fujian, Gansu, Guangdong, Guizhou, Hainan, Hebei, Heilongjiang, Henan, Hubei, Hunan, Jiangsu, Jiangxi, Jilin, Liaoning, Qinghai, Shaanxi, Shandong, Shanxi, Sichuan, Taiwan (claimed by China), Yunnan, Zhejiang (23 provinces) •Guangxi, Nei Mongol, Ningxia, Xinjiang, Xizang (5 autonomous regions) •Beijing, Chongqing, Shanghai, Tianjin (4 municipalities)
Special Administrative Regions: Hong Kong, Macau
Ethnic/Racial Groups: Han, and 55 other groups, including Tibetan and Mongol
Religions: officially atheist; Taoism, Buddhism, Islam
Languages: Mandarin, Yue (Cantonese), other dialects and minority languages
Currency: yuan
Current President: Hu Jintao
Cities (capital, largest, or with at least 3.5 million people):
   Shanghai: •located in eastern China •city in the Shanghai municipality •most populated city in China (16,575,000 people)
   Beijing: •located in eastern China •city in the Beijing municipality •capital of China (12,214,000 people)
   Wuhan: •located in eastern China •city in the Hubei province •a major port on the Chang Jiang (Yangtze River)
   Shenzhen: •located in southern China •city in the Guangdong province
   Chengdu: •located in central China •city in the Sichuan province
   Tianjin: •located in eastern China •city in the Tianjin municipality
   Shenyang: •located in northeastern China •city in the Liaoning province
   Chongqing: •located in southern China •city in the Chongqing municipality •a major port on the Chang Jiang (Yangtze River)
   Guangzhou: •located in southeastern China •city in the Guangdong province
   Harbin: •located in northeastern China •city in the Heilongjiang province
   Xian: •located in central China •city in the Shanxi province
   Nanjing: •located in eastern China •city in the Jiangsu province •a major port on the Chang Jiang (Yangtze River)

Taiyuan: • located in central China city in the Shanxi province
Changchun: • located in northeastern China • city in the Jilin province
Zhengzhou: • located in eastern China • city in the Henan province
Jiulong: • located in southeastern China • city in Hong Kong
Jinan: • located in eastern China • city in the Shandong province
Dalian: • located in northeastern China • city in the Liaoning province
Qingdao: • located in eastern China • city in the Shandong province

ENVIRONMENTAL/ECONOMIC:
Climate: tropical in the south; subarctic in the north
Natural Resources: coal, iron ore, petroleum, natural gas, mercury
Agricultural Products: rice, wheat, potatoes, sorghum, pork
Major Exports: machinery and equipment, textiles and clothing, footwear, toys and sporting goods, mineral fuels
Natural Hazards: typhoons, floods, tsunamis, earthquakes, droughts, land subsidence

COLOMBIA
Country Name: Republic of Colombia
Continent: South America
Area: 440,831 sq mi / 1,141,748 sq km
Population: 46,871,000
Population Density: 106 people per sq mi / 41 people per sq km
Capital: Bogota

PHYSICAL:
Highest Points:
   Pico Cristobal Colon 18,947 ft / 5,775 m
   Pico Simon Bolivar 18,947 ft / 5,775 m
Lowest Point: Pacific Ocean 0 ft / 0 m (sea level)
Mountain Ranges:
   Andes: • cover much of western and central Colombia • extend into Ecuador
   Sierra Nevada de Santa Marta: • located in northern Colombia
Peaks (minimum elevation of 8,000 ft / 2,400 m):
   Pico Cristobal Colon: • located in the Sierra Nevada de Santa Marta • one of the highest points in Colombia
   Pico Simon Bolivar: • located in the Sierra Nevada de Santa Marta • one of the highest points in Colombia
Grasslands/Prairies:
   Llanos: • located in eastern Colombia • extend into Venezuela
Rain Forests:
   Amazon: • located in southeastern Colombia • extends into Ecuador, Peru, Brazil, and Venezuela • largest rain forest in the world

Oceans:
   Pacific: •forms part of the western coast of Colombia
Seas:
   Caribbean: •forms most of the northern coast of Colombia •feeds into the Atlantic Ocean
Gulfs:
   Gulf of Venezuela: •forms part of the northeastern coast of Colombia •feeds into the Caribbean Sea
Rivers:
   Cauca: •tributary of the Magdalena River •has its source in the Andes
   Guaviare: •tributary of the Orinoco River •has its source in the Andes
   Magdalena: •has its mouth in the Caribbean Sea •has its source in the Andes
   Meta: •tributary of the Orinoco River •has its source in the Andes •forms part of the Colombia-Venezuela border
   Negro: •forms part of the Colombia-Venezuela border •extends into Brazil
   Orinoco: •forms part of the Colombia-Venezuela border •extends into Venezuela
   Putumayo: •has its source in the Andes •forms most of the Colombia-Peru border •forms part of the Colombia-Ecuador border
Peninsulas:
   Guajira: •bordered by the Caribbean Sea and the Gulf of Venezuela •shares the border with Venezuela

POLITICAL:
Independence: July 20, 1810 (from Spain)

Bordering Countries: Venezuela, Brazil, Peru, Ecuador, Panama (5)
Administrative Divisions: Amazonas, Antioquia, Arauca, Atlantico, Bolivar, Boyaca, Caldas, Caqueta, Casanare, Cauca, Cesar, Choco, Cordoba, Cundinamarca, Guainia, Guajira, Guaviare, Huila, Magdalena, Meta, Narino, Norte de Santander, Putumayo, Quindio, Risaralda, San Andres y Providencia, Santander, Sucre, Tolima, Valle del Cauca, Vaupes, Vichada (32 departments) ●Distrito Capital de Bogota (1 capital district)
Ethnic/Racial Groups: mestizo, European, Creole, black, Amerindian
Religion: Christianity (mainly Roman Catholic)
Language: Spanish
Currency: Colombian peso
Current President: Juan Manuel Santos Calderon
Cities (capital, largest, or with at least a million people):
Bogota: ●located in central Colombia ●city in the Distrito Capital (capital district) ●capital of Colombia ●most populated city in Colombia (8,262,000 people) ●located in a valley in the Andes
Cali: ●located in western Colombia ●city in the Valle de Cauca department ●located in a valley in the Andes ●chief port on the Cauca River
Medellin: ●located in western Colombia ●city in the Antioquia department ●located in a valley in the Andes
Barranquilla: ●located in northern Colombia ●city in the Atlantico department ●a major port on the Magdalena River and the Caribbean Sea

ENVIRONMENTAL/ECONOMIC:
Climate: tropical along the coast and eastern plains; cooler in highlands
Natural Resources: petroleum, natural gas, coal, iron ore, nickel
Agricultural Products: coffee, cut flowers, bananas, rice, forest products
Major Exports: petroleum, coffee, coal, apparel, bananas
Natural Hazards: volcanoes, earthquakes, droughts

COMOROS
Country Name: Union of the Comoros
Continent: Africa
Area: 719 sq mi / 1,862 sq km
Population: 754,000
Population Density: 1,049 people per sq mi / 405 people per sq km
Capital: Moroni

PHYSICAL:
Highest Point: Kartala 7,743 ft / 2,360 m
Lowest Point: Indian Ocean 0 ft / 0 m (sea level)
Oceans:
Indian: ●forms part of the coast of Comoros

Channels:
Mozambique: •forms part of the coast of Comoros •separates Madagascar and Comoros from mainland Africa
Islands:
Grande Comore: •largest island in Comoros •bordered by the Indian Ocean and the Mozambique Channel

POLITICAL:
Independence: July 6, 1975 (from France)
Administrative Divisions: Anjouan, Grande Comore, Moheli (3 islands)
Ethnic/Racial Groups: Antalote, Cafre, Makoa, Oimatsaha, Sakalava
Religions: Islam (predominantly Sunni), Christianity (mainly Roman Catholic)
Languages: Arabic, French, Shikomoro
Currency: Comoran franc
Current President: Ikililou Dhoinine
Cities (capital, largest, or with at least a million people):
Moroni: •located on Grande Comore Island •city in the Grande Comore Island division •capital of Comoros •most populated city in Comoros (49,000 people) •chief port on the Mozambique Channel

ENVIRONMENTAL/ECONOMIC:
Climate: tropical marine with a rainy season
Agricultural Products: vanilla, cloves, perfume essences, copra
Major Exports: vanilla, ylang-ylang, cloves, perfume oil, copra
Natural Hazards: cyclones, volcanoes

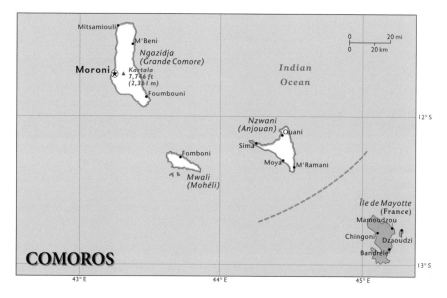

CONGO, REPUBLIC OF THE

Country Name: Republic of the Congo
Continent: Africa
Area: 132,047 sq mi / 342,000 sq km
Population: 4,144,000
Population Density: 31 people per sq mi / 12 people per sq km
Capital: Brazzaville

PHYSICAL:
Highest Point: Mount Berongou 2,963 ft / 903 m
Lowest Point: Atlantic Ocean 0 ft / 0 m (sea level)
Rain Forests:
  Congo: •covers much of northern and central Republic of the Congo •extends into Democratic Republic of the Congo and Gabon
Oceans:
  Atlantic: •forms the southwestern coast of the Republic of the Congo
Rivers:
  Congo: •forms part of the Republic of the Congo-Democratic Republic of the Congo border •extends into Democratic Republic of the Congo •tenth longest river in the world
  Ubangi (Oubangui): •tributary of the Congo River •forms part of the Republic of the Congo-Democratic Republic of the Congo border

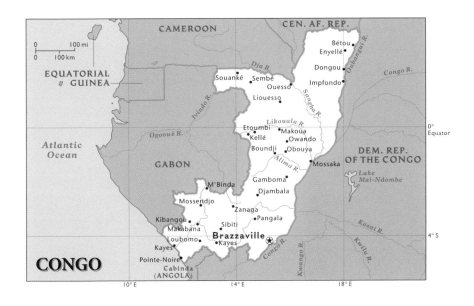

POLITICAL:
Independence: August 15, 1960 (from France)
Former Names: Middle Congo, Congo/Brazzaville, Congo
Bordering Countries: Gabon, Cameroon, Central African Republic, Democratic Republic of the Congo, Angola (5)
States/Provinces: Bouenza, Cuvette, Cuvette-Ouest, Kouilou, Lekoumou, Likouala, Niari, Plateaux, Pool, Sangha (10 regions) •Brazzaville (1 commune)
Ethnic/Racial Groups: Kongo, Sangha, Teke, M'Bochi
Religions: Christianity, animism, Islam
Languages: French, Lingala, Monokutuba, many local dialects
Currency: Communaute Finaciere Africaine franc
Current President: Denis Sassou-Nguesso
Cities (capital, largest, or with at least a million people):
    Brazzaville: •located in southeastern Republic of the Congo •city in the Pool region •Brazzaville commune is located in center of the city •capital of Republic of the Congo •most populated city in Republic of the Congo (1,292,000) •chief port on the Congo River

ENVIRONMENTAL/ECONOMIC:
Climate: tropical with rainy/dry seasons; hot and humid
Natural Resources: petroleum, timber, potash, lead, zinc
Agricultural Products: cassava (tapioca), sugar, rice, corn, forest products
Major Exports: petroleum, lumber, plywood, sugar, cacao
Natural Disasters: floods

CONGO, DEMOCRATIC REPUBLIC OF THE
Country Name: Democratic Republic of the Congo
Continent: Africa
Area: 905,365 sq mi / 2,344,885 sq km
Population: 67,823,000
Population Density: 75 people per sq mi / 29 people per sq km
Capital: Kinshasa

PHYSICAL:
Highest Point: Margherita Peak (on Mount Stanley) 16,765 ft / 5,110 m
Lowest Point: Atlantic Ocean 0 ft / 0 m (sea level)
Mountain Ranges:
    Bleus: •located in northeastern Democratic Republic of the Congo •extend into Uganda
    Mitumba: •cover much of eastern Democratic Republic of the Congo •extend into Uganda, Rwanda, Burundi, and Tanzania
Peaks (minimum elevation of 8,000 ft / 2,400 m):
    Mount Stanley: •located in the Bleus Mountains •on the Democratic Republic of the Congo-Uganda border •includes Margherita Peak, the highest point in Democratic Republic of the Congo

Valleys:
  Great Rift: ●located in eastern Democratic Republic of the Congo ●extends into Uganda, Rwanda, Burundi, Tanzania, and Zambia
Rain Forests:
  Congo: ●covers most of the Democratic Republic of the Congo ●extends into the Republic of the Congo
Oceans:
  Atlantic: ●forms the western coast of Democratic Republic of the Congo
Lakes:
  Lake Albert: ●located in northeastern Democratic Republic of the Congo ●fed by the Victoria Nile River ●on the Democratic Republic of the Congo-Uganda border ●located on the edge of the Bleus Mountains in the Great Rift Valley
  Lake Edward: ●located in northeastern Democratic Republic of the Congo ●on the Democratic Republic of the Congo-Uganda border ●located on the edge of the Bleus Mountains in the Great Rift Valley
  Lake Kivu: ●located in eastern Democratic Republic of the Congo ●on the Democratic Republic of the Congo-Rwanda border with ●located on the edge of the Mitumba Mountains in the Great Rift Valley
  Lake Mweru: ●located in southeastern Democratic Republic of the Congo ●on the Democratic Republic of the Congo–Zambia border ●located on the edge of the Mitumba Mountains
  Lake Tanganyika: ●located in southeastern Democratic Republic of the Congo ●on the border with Burundi, Tanzania, and Zambia ●located on the edge of the Mitumba Mountains in the Great Rift Valley ●sixth largest lake in the world

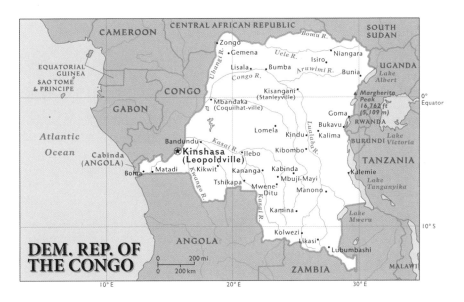

Rivers:
  Congo: •has its mouth in the Atlantic Ocean •has its source in the Mitumba Mountains •forms part of Democratic Republic of the Congo's border with the Republic of the Congo and Angola •tenth longest river in the world
  Kasai: •tributary of the Kwa River •forms part of the Democratic Republic of the Congo–Angola border •extends into Angola
  Kwa: •tributary of the Congo River
  Kwango: •tributary of the Kwa River •forms part of the Democratic Republic of the Congo–Angola border •extends into Angola
  Kwilu: •tributary of the Kwango River •extends into Angola
  Lualaba: •tributary of the Congo River •has its source in the Mitumba Mountains
  Ubangi: •tributary of the Congo River •has its source in the Bleus Mountains •forms part of the Democratic Republic of the Congo's border with the Republic of the Congo and Central African Republic
Deltas:
  Congo River Delta: •mouth of the Congo River •feeds into the Atlantic Ocean •borders Democratic Republic of the Congo and Angola
Waterfalls:
  Stanley (or Boyoma): •located in northern Democratic Republic of the Congo •part of the Lualaba River

POLITICAL:
Independence: June 30, 1960 (from Belgium)
Former Names: Congo Free State, Belgian Congo, Katanga, Congo/Leopoldville, Congo/Kinshasa, Zaire
Bordering Countries: Angola, Republic of the Congo, Zambia, Central African Republic, Uganda, South Sudan, Tanzania, Burundi, Rwanda (9)
Administrative Divisions: Bandundu, Bas-Congo, Equator, Kasai-Occidental, Kasai-Oriental, Katanga, Maniema, Nord-Kivu, Orientale, Sud-Kivu (10 provinces) •Kinshasa (1 city)
Ethnic/Racial Groups: more than 200 African groups, mainly Bantu (Mongo, Luba, Kongo) and Hamitic (Mangbetu-Azande)
Religions: Christianity (Roman Catholic, Protestant), Kimbanguist (indigenous), Islam, traditional
Languages: French, Lingala, Kingwana, Kikongo, Tshiluba
Currency: Congolese franc
Current President: Joseph Kabila
Cities (capital, largest, or with at least a million people):
  Kinshasa: •located in western Democratic Republic of the Congo •city in the Kinshasa city division •capital of the Democratic Republic of the Congo •most populated city in Democratic Republic of the Congo (8,401,000 people) •chief port on the Congo River
  Lubumbashi: •located in southeastern Democratic Republic of the Congo •city in the Katanga province

Mbuji-Mayi: •located in south-central Democratic Republic of the Congo •city in the Kasai-Oriental province

ENVIRONMENTAL/ECONOMIC:
Climate: tropical; hot and humid in the equatorial basin; cooler and drier in the southern highlands; cooler and wetter in the eastern highlands; wet/dry seasons
Natural Resources: cobalt, copper, cadmium, petroleum, industrial and gem diamonds
Agricultural Products: coffee, sugar, palm oil, rubber, wood products
Major Exports: diamonds, copper, crude oil, coffee, cobalt
Natural Hazards: droughts, floods, volcanoes

COSTA RICA
Country Name: Republic of Costa Rica
Continent: North America
Area: 19,730 sq mi / 51,100 sq km
Population: 4,727,000
Population Density: 240 people per sq mi / 93 people per sq km
Capital: San Jose

PHYSICAL:
Highest Point: Cerro Chirripo 12,500 ft / 3,810 m
Lowest Point: Pacific Ocean 0 ft / 0 m (sea level)
Mountain Ranges:
   Cordillera de Talamanca: •covers much of northwestern and southeastern Costa Rica •extends from Costa Rica into Panama
Peaks (minimum elevation of 8,000 ft / 2,400 m):
   Cerro Chirripo: •located in the Cordillera de Talamanca •highest point in Costa Rica
Oceans:
   Pacific: •forms most of the western and southern coasts of Costa Rica
Seas:
   Caribbean: •forms part of the eastern coast of Costa Rica •feeds into the Atlantic Ocean

POLITICAL:
Independence: September 15, 1821 (from Spain)
Bordering Countries: Panama, Nicaragua (2)
Administrative Divisions: Alajuela, Cartago, Guanacaste, Heredia, Limon, Puntarenas, San Jose (7 provinces)
Ethnic/Racial Groups: European, mestizo, black, Amerindian, Chinese
Religion: Christianity (Roman Catholic, Evangelical)
Languages: Spanish, English
Currency: Costa Rican colon
Current President: Laura Chinchilla Miranda
Cities (capital, largest, or with at least a million people):

San Jose: •located in central Costa Rica •city in the San Jose province •capital of Costa Rica •most populated city in Costa Rica (1,416,000 people) •located in a valley near the Cordillera de Talamanca

ENVIRONMENTAL/ECONOMIC:
Climate: tropical to subtropical with rainy/dry seasons; cooler in the highlands
Natural Resources: hydropower
Agricultural Products: coffee, pineapples, bananas, sugar, beef, timber
Major Exports: coffee, bananas, sugar, pineapples, textiles
Natural Hazards: earthquakes, hurricanes, floods, landslides, volcanoes

COTE D'IVOIRE

Country Name: Republic of Cote d'Ivoire
Continent: Africa
Area: 124,503 sq mi / 322,462 sq km
Population: 22,621,000
Population Density: 182 people per sq mi / 70 people per sq km
Capitals: Abidjan (administrative capital), Yamoussoukro (legislative capital)

PHYSICAL:
Highest Point: Mont Nimba 5,748 ft / 1,752 m
Lowest Point: Gulf of Guinea 0 ft / 0 m (sea level)
Gulfs:
   Gulf of Guinea: •forms the southern coast of Cote d'Ivoire •feeds into the Atlantic Ocean

Lakes:
Lake Kossou: •located in central Cote d'Ivoire

POLITICAL:
Independence: August 7, 1960 (from France)
Bordering Countries: Liberia, Ghana, Guinea, Burkina Faso, Mali (5)
Administrative Divisions: Agneby, Bafing, Bas-Sassandra, Denguele, Dix-Huit Montagnes, Fromager, Haut-Sassandra, Lacs, Lagunes, Marahoue, Moyen-Cavally, Moyen-Comoe, N'zi-Comoe, Savanes, Sud-Bandama, Sud-Comoe, Vallee du Bandama, Worodougou, Zanzan (19 regions)
Ethnic/Racial Groups: Akan, Gur, Northern Mandes, Krous, Southern Mandes
Religions: Christianity, Islam, indigenous beliefs
Languages: French, Dioula, 60 native dialects
Currency: Communaute Finaciere Africaine franc
Current President: Alassane Ouattara
Cities (capital, largest, or with at least a million people):
Abidjan: •located in southeastern Cote d'Ivoire •city in the Abidjan department •administrative capital of Cote d'Ivoire •most populated city in Cote d'Ivoire (4,009,000 people) •chief port on the Gulf of Guinea
Yamoussoukro: •located in central Cote d'Ivoire •legislative capital of Cote d'Ivoire (808,000 people)

ENVIRONMENTAL/ECONOMIC:
Climate: tropical along the coast, semi-arid in the far north
Natural Resources: petroleum, natural gas, diamonds, manganese, iron ore

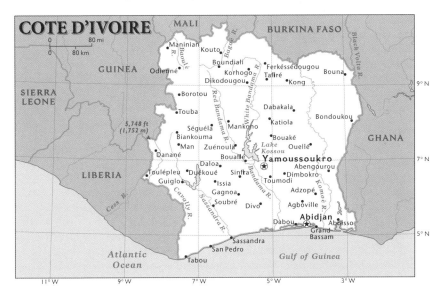

Agricultural Products: coffee, cacao, bananas, palm kernels, timber
Major Exports: cacao, coffee, timber, petroleum, cotton
Natural Hazards: tall waves, flooding

CROATIA
Country Name: Republic of Croatia
Continent: Europe
Area: 21,831 sq mi / 56,542 sq km
Population: 4,405,000
Population Density: 202 people per sq mi / 78 people per sq km
Capital: Zagreb

PHYSICAL:
Highest Point: Dinara 6,007 ft / 1,831 m
Lowest Point: Adriatic Sea 0 ft / 0 m (sea level)
Mountain Ranges:
  Dinaric Alps: •cover much of western Croatia •extend into Bosnia and Herzegovina and into Serbia and Montenegro
Grasslands/Prairies:
  Great Alfold: •located in northeastern Croatia •extends into Serbia and Montenegro and into Hungary
Seas:
  Adriatic: •forms most of the western and southern coasts of Croatia •feeds into the Ionian Sea
Rivers:
  Danube: •forms part of the Croatia–Serbia and Montenegro border •extends into Hungary and into Serbia and Montenegro
  Drava: •tributary of the Danube River •forms part of the Croatia-Hungary border •extends into Slovenia
  Sava: •forms part of the Croatia–Bosnia and Herzegovina border •extends into Slovenia and into Serbia and Montenegro
Regions:
  Dalmatia: •mountainous coastal region in southern Croatia •located east of the Adriatic Sea in the Dinaric Alps

POLITICAL:
Independence: June 25, 1991 (from Yugoslavia)
Former Names: People's Republic of Croatia, Socialist Republic of Croatia
Bordering Countries: Bosnia and Herzegovina, Slovenia, Hungary, Serbia and Montenegro (4)
Administrative Divisions: Bjelovarsko-Bilogorska, Brodsko-Posavska, Dubrovacko-Neretvanska, Istarska, Karlovacka, Koprivnicko-Krizevacka, Krapinsko-Zagorska, Licko-Senjska, Medimurska Zupanija, Osjecko-Baranjska, Pozesko-Slavonska,

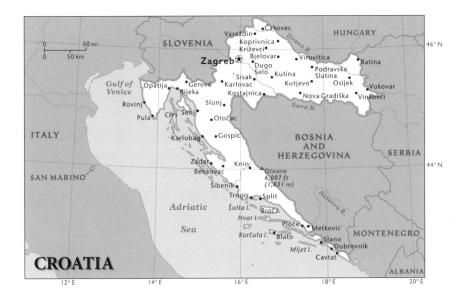

Primorsko-Goranska, Sibensko-Kninska, Sisacko-Moslavacka, Splitsko-Dalmatinska, Varazdinska, Viroviticko-Podravska, Vukovarsko-Srijemska, Zadarska, Zagrebacka (20 counties) •Zagreb (1 city)

Regions:
  Slavonia: •region located in central and eastern Croatia
Ethnic/Racial Groups: Croat, Serb, Bosniak
Religion: Christianity (Roman Catholic, Orthodox)
Language: Croatian
Currency: kuna
Current President: Ivo Josipovic
Cities (capital, largest, or with at least a million people):
  Zagreb: •located in northern Croatia •city in the Zagreb city division •capital of Croatia •most populated city in Croatia (685,000 people) •chief port on the Sava River

ENVIRONMENTAL/ECONOMIC:
Climate: Mediterranean and continental with hot summers/cold winters
Natural Resources: oil, coal, bauxite, low-grade iron ore, calcium
Agricultural Products: wheat, corn, sugar beets, sunflower seed, livestock
Major Exports: transport equipment, textiles, chemicals, foodstuffs, fuels
Natural Hazards: earthquakes

CUBA

Country Name: Republic of Cuba
Continent: North America
Area: 42,803 sq mi / 110,860 sq km
Population: 11,240,000
Population Density: 263 people per sq mi / 101 people per sq km
Capital: Havana

PHYSICAL:
Highest Point: Pico Turquino 6,578 ft / 2,005 m
Lowest Point: Caribbean Sea 0 ft / 0 m (sea level)
Oceans:
  Atlantic: •forms part of the northern coast of Cuba
Seas:
  Caribbean: •forms most of the southern coast of Cuba •feeds into the Atlantic Ocean
Gulfs:
  Gulf of Mexico: •forms part of the northwestern coast of Cuba •feeds into the Atlantic Ocean
Straits:
  Straits of Florida: •forms part of the northern coast of Cuba •separates Cuba from the United States •connects the Gulf of Mexico and the Atlantic Ocean
Channels:
  Yucatan: •forms part of the western coast of Cuba •separates Cuba from Mexico •connects the Gulf of Mexico and the Caribbean Sea
Passages:
  Windward: •forms part of the southeastern coast of Cuba •separates Cuba from the island of Hispaniola

POLITICAL:
Independence: May 20, 1902 (from Spain)
Administrative Divisions: Artemisa, Camaguey, Ciego de Avila, Cienfuegos, Granma, Guantanamo, Holguin, La Habana, Las Tunas, Mayabeque, Matanzas, Pinar del Rio, Sancti Spiritus, Santiago de Cuba, Villa Clara (15 provinces) •Isla de la Juventud (1 special municipality)
Ethnic/Racial Groups: Creole, white, black
Religions: Christianity (Roman Catholic, Protestant, Jehovah's Witness), Judaism, Santeria
Language: Spanish
Currency: Cuban peso
Current President: Raul Castro Ruz
Cities (capital, largest, or with at least a million people):
  Havana: •located in northwestern Cuba •city in the La Habana province •capital of Cuba •most populated city in Cuba (2,140,000 people) •chief port on the Gulf of Mexico

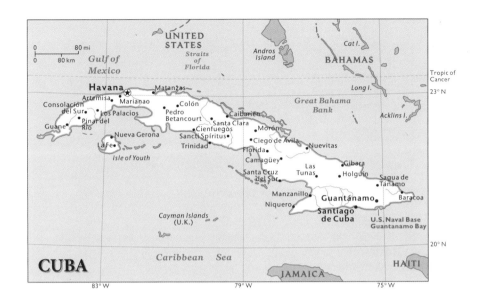

ENVIRONMENTAL/ECONOMIC:
Climate: tropical, moderated by trade winds; dry/rainy season
Natural Resources: cobalt, nickel, iron ore, chromium, copper
Agricultural Products: sugar, tobacco, citrus, coffee, livestock
Major Exports: sugar, nickel, tobacco, fish, medical products
Natural Hazards: hurricanes, droughts

CYPRUS
Country Name: Republic of Cyprus
Continent: Europe
Area: 3,572 sq mi / 9,251 sq km
Population: 1,107,000
Population Density: 310 people per sq mi / 120 people per sq km
Capital: Nicosia

PHYSICAL:
Highest Point: Mount Olympus 6,401 ft / 1,951 m
Lowest Point: Mediterranean Sea 0 ft / 0 m
Seas:
    Mediterranean: •forms the entire coast of Cyprus •feeds into the Atlantic Ocean

POLITICAL:
Independence: August 16, 1960 (from the United Kingdom)
Administrative Divisions: Famagusta, Kyrenia, Larnaca, Limassol, Nicosia, Paphos (6 districts)
Regions: Northern Cyprus (occupied by Turkey since 1974)
Ethnic/Racial Groups: Greek, Turkish
Religions: Christianity (primarily Greek Orthodox), Islam
Languages: Greek, Turkish, English
Currency: euro, Turkish new lira
Current President: Demetris Christofias
Cities (capital, largest, or with at least a million people):
  Nicosia: •located in central Cyprus •city in the Nicosia district •capital of Cyprus •most populated city in Cyprus (240,000 people)

ENVIRONMENTAL/ECONOMIC:
Climate: temperate, Mediterranean; hot and dry summers/cool winters
Natural Resources: copper, pyrites, asbestos, gypsum, timber
Agricultural Products: potatoes, citrus, vegetables, barley

Major Exports: citrus, potatoes, pharmaceuticals, cement (Greek area); citrus, potatoes, textiles (Turkish area)
Natural Hazards: earthquakes, droughts

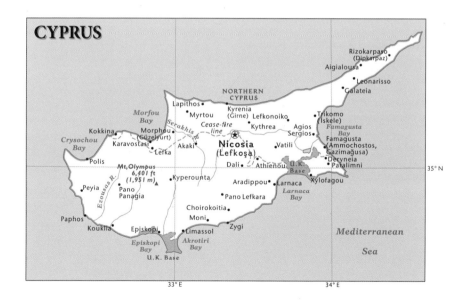

CZECH REPUBLIC

Country Name: Czech Republic
Continent: Europe
Area: 30,450 sq mi / 78,866 sq km
Population: 10,546,000
Population Density: 346 people per sq mi / 134 people per sq km
Capital: Prague

PHYSICAL:
Highest Point: Snezka 5,256 ft / 1,602 m
Lowest Point: Elbe River 377 ft / 115 m
Mountain Ranges:
- Carpathian: •located in southeastern Czech Republic •extend into Slovakia and Poland
- Erzgebirge: •located in northwestern Czech Republic •extend into Germany
- Sudeten: •cover much of northern Czech Republic •extend into Poland

Rivers:
- Elbe: •has its source in the Sudeten Mountains •extends into Germany
- Morava: •has its source in the Sudeten Mountains •forms part of the Czech Republic–Slovakia border
- Oder: •has its source in the Sudeten Mountains •extends into Poland
- Vltava: •tributary of the Elbe River •has its source in southwestern Czech Republic

POLITICAL:
Independence: January 1, 1993 (from Czechoslovakia)
Bordering Countries: Poland, Germany, Austria, Slovakia (4)
Administrative Divisions: Jihocesky, Jihomoravsky, Karlovarsky, Kralovehradecky, Liberecky, Moravskoslezsky, Olomoucky, Pardubicky, Plzensky, Stredocesky, Ustecky, Vysocina, Zlinsky (13 regions) •Prague (1 capital city)
Ethnic/Racial Groups: Czech, Moravian, Slovak
Religion: Christianity (Roman Catholic, Protestant), largest percent is atheist
Language: Czech
Currency: Czech koruna
Current President: Vaclav Klaus
Cities (capital, largest, or with at least a million people):
- Prague: •located in central Czech Republic •city in the Praha capital city division •capital of the Czech Republic •most populated city in the Czech Republic (1,162,000 people) •chief port on the Vltava River

ENVIRONMENTAL/ECONOMIC:
Climate: temperate with cool summers/cold winters
Natural Resources: hard coal, soft coal, kaolin, clay, graphite
Agricultural Products: wheat, potatoes, sugar beets, hops, pigs
Major Exports: machinery and equipment, intermediate manufactures, chemicals, raw materials, fuel
Natural Hazards: floods

DENMARK

Country Name: Kingdom of Denmark
Continent: Europe
Area: 16,640 sq mi / 43,098 sq km
Population: 5,574,000
Population Density: 335 people per sq mi / 129 people per sq km
Capital: Copenhagen

PHYSICAL:

Highest Point: Mollehoj/Ejer Bavnehoj 561 ft / 171 m
Lowest Point: Lammefjord 23 ft / 7 m below sea level
Seas:
  North: • forms most of the western coast of Denmark • feeds into the Atlantic Ocean
  Baltic: • forms part of the southeastern coast of Denmark • feeds into the Kattegat
Gulfs:
  Kattegat: • forms part of the eastern and northern coasts of Denmark • feeds into the North Sea
Bays:
  Alborg: • forms part of the northeastern coast of Denmark • feeds into the Kattegat
  Kieler (Kiel): • forms part of the southern coast of Denmark • feeds into the Kattegat
  Mecklenburger: • forms part of the southeastern coast of Denmark • feeds into the Baltic Sea
Straits:
  Store Baelt: • forms part of the southern coast of Denmark • separates the island of Fyn from the island of Sjaelland • connects Kiel Bay and the Kattegat
Channels:
  Skagerrak: • forms part of the northern coast of Denmark • separates the Jutland Peninsula from Norway • connects the Kattegat and the North Sea
Passages:
  Oresund: • forms part of the eastern coast of Denmark • separates the island of Sjaelland from Sweden • connects the Kattegat and the Baltic Sea
Peninsulas:
  Jutland: • bordered by the North Sea, the Skagerrak, and the Kattegat
Islands:
  Bornholm: • island off the southern coast of Sweden • surrounded by the Baltic Sea
  Fyn: • island off of the southeastern coast of Jutland Peninsula • bordered by the Store Baelt and Kiel Bay

Lolland: •island located south of Sjaelland •bordered by Kiel Bay, Mecklenburger Bay, and the Baltic Sea

Sjaelland: •island located east of Fyn •bordered by the Kattegat, Store Baelt, Oresund, and the Baltic Sea

POLITICAL:
Independence: organized in the 10th century; constitution established on June 5, 1849
Bordering Country: Germany (1)
Administrative Divisions: Hovedstaden, Midtjylland, Nordjylland, Sjaelland, Syddanmark (5 regions)
External Territories: Faroe Islands, Greenland (2)
Ethnic/Racial Groups: Danish, German
Religion: Christianity (primarily Evangelical Lutheran)
Languages: Danish, Faroese, Greenlandic, German
Currency: Danish krone
Current Prime Minister: Helle Thorning-Schmidt
Cities (capital, largest, or with at least a million people):
Copenhagen: •located in eastern Denmark •city in Kobenhavn county •Kobenhavns borough is located in the center of the city •capital of Denmark •most populated city in Denmark (1,174,000 people) •chief port on the Oresund and the Baltic Sea

ENVIRONMENTAL/ECONOMIC:
Climate: temperate with mild winters/cool summers
Natural Resources: petroleum, natural gas, fish, salt, limestone
Agricultural Products: barley, wheat, potatoes, sugar beets, pork
Major Exports: machinery and instruments, meat and meat products, dairy products, fish, chemicals
Natural Hazards: floods

DJIBOUTI
Country Name: Republic of Djibouti
Continent: Africa
Area: 8,958 sq mi / 23,200 sq km
Population: 906,000
Population Density: 101 people per sq mi / 39 people per sq km
Capital: Djibouti

PHYSICAL:
Highest Point: Mous alli 6,631 ft / 2,021 m
Lowest Point: Lake Assal 512 ft / 156 m below sea level
Valleys:
  Great Rift: •covers an area from southwest to northeast Djibouti •extends into Ethiopia
Depressions:
  Lake Assal: •located in the Great Rift Valley •lowest point in Africa
Gulfs:
  Gulf of Aden: •forms part of the eastern coast of Djibouti •feeds into the Arabian Sea
  Gulf of Tadjoura: •inlet of the Gulf of Aden between Tadjoura and Djibouti city
Straits:
  Bab al Mandab: •forms the northeastern coast of Djibouti •separates Djibouti from Yemen •connects the Red Sea and the Gulf of Aden
Regions:
  Denakil: •lowland, geothermal region that covers much of western Djibouti •extends into Ethiopia and Eritrea •located north of the Great Rift Valley

POLITICAL:
Independence: June 27, 1977 (from France)
Former Names: French Territory of the Afars and Issas, French Somaliland
Bordering Countries: Ethiopia, Eritrea, Somalia (3)
Administrative Divisions: Ali Sabieh, Arta, Dikhil, Djibouti, Obock, Tadjoura (6 districts)
Ethnic/Racial Groups: Somali, Afar
Religions: Islam, Christianity
Languages: French, Arabic, Somali, Afar
Currency: Djiboutian franc

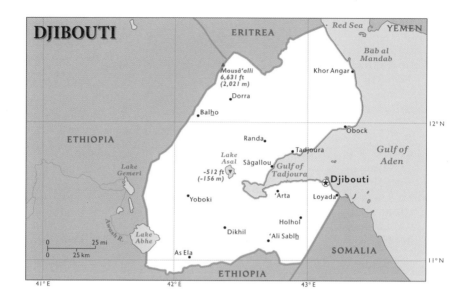

Current President: Ismail Omar Guelleh
Cities (capital, largest, or with at least a million people):
  Djibouti: •located in eastern Djibouti •city in the Djibouti district •capital of Djibouti •most populated city in Djibouti (567,000 people) •chief port on the Gulf of Aden

ENVIRONMENTAL/ECONOMIC:
Climate: desert, dry
Natural Resources: geothermal areas
Agricultural Products: fruits, vegetables, goats
Major Exports: re-exports, hides and skins, coffee (in transit)
Natural Disasters: earthquakes, droughts, cyclones, flash floods

DOMINICA
Country Name: Commonwealth of Dominica
Continent: North America
Area: 290 sq mi / 751 sq km
Population: 73,000
Population Density: 252 people per sq mi / 97 people per sq km
Capital: Roseau

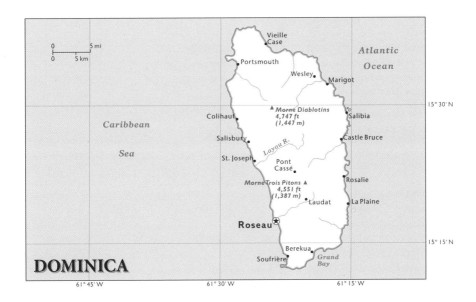

DOMINICA

PHYSICAL:
Highest Point: Morne Diablotins 4,747 ft / 1,447 m
Lowest Point: Caribbean Sea 0 ft / 0 m (sea level)
Oceans:
   Atlantic: • forms the eastern coast of Dominica
Seas:
   Caribbean: • forms the western coast of Dominica • feeds into the Atlantic Ocean

POLITICAL:
Independence: November 3, 1978 (from the United Kingdom)
Administrative Divisions: Saint Andrew, Saint David, Saint George, Saint John, Saint Joseph, Saint Luke, Saint Mark, Saint Patrick, Saint Paul, Saint Peter (10 parishes)
Ethnic/Racial Groups: black, Creole, white, Carib Amerindian
Religion: Christianity (Roman Catholic, Protestant)
Languages: English, French patois
Currency: East Caribbean dollar
Current President: Nicholas J. O. Liverpool
Cities (capital, largest, or with at least a million people):
   Roseau: • located in southwestern Dominica • city in the Saint George parish • capital of Dominica • most populated city in Dominica (14,000 people) • chief port on the Caribbean Sea

ENVIRONMENTAL/ECONOMIC:
Climate: tropical, moderated by trade winds; heavy rains
Natural Resources: timber, hydropower, arable land
Agricultural Products: bananas, citrus, mangoes, root crops
Major Exports: bananas, soap, bay oil, vegetables, grapefruit
Natural Hazards: flash floods, hurricanes

DOMINICAN REPUBLIC
Country Name: Dominican Republic
Continent: North America
Area: 18,704 sq mi / 48,442 sq km
Population: 10,010,000
Population Density: 535 people per sq mi / 207 people per sq km
Capital: Santo Domingo

PHYSICAL:
Highest Point: Pico Duarte 10,417 ft / 3,175 m
Lowest Point: Lake Enriquillo 151 ft / 46 m below sea level
Peaks (minimum elevation of 8,000 ft / 2,400 m):
   Pico Duarte: •located in western Dominican Republic •highest point in Dominican Republic •highest point in the Caribbean
Oceans:
   Atlantic: •forms the northern coast of Dominican Republic
Seas:
   Caribbean: •forms the southern coast of Dominican Republic •feeds into the Atlantic Ocean
Passages:
   Mona: •forms part of the eastern coast of Dominican Republic •separates the island of Hispaniola from Puerto Rico (U.S. commonwealth) •connects the Caribbean Sea and the Atlantic Ocean
Islands:
   Hispaniola: •Dominican Republic takes up the eastern portion of this island •bordered by the Atlantic Ocean and the Caribbean Sea •Haiti takes up the western portion of the island

POLITICAL:
Independence: February 27, 1844 (from Haiti)
Bordering Countries: Haiti (1)
Administrative Divisions: Azua, Baoruco, Barahona, Dajabon, Duarte, Elias Pina, El Seibo, Espaillat, Hato Mayor, Independencia, La Altagracia, La Romana, La Vega, Maria Trinidad Sanchez, Monsenor Nouel, Monte Cristi, Monte Plata, Pedernales, Peravia, Puerto Plata, Salcedo, Samana, Sanchez Ramirez, San Cristobal, San Jose de Ocoa, San Juan, San Pedro de Macoris, Santiago, Santiago Rodriguez, Santo Domingo, Valverde (31 provinces) •Distrito Nacional (1 district)

Ethnic/Racial Groups: Creole, white, black
Religion: Christianity (primarily Roman Catholic)
Language: Spanish
Currency: Dominican peso
Current President: Leonel Fernandez Reyna
Cities (capital, largest, or with at least a million people):
  Santo Domingo: •located in southern Dominican Republic •city in the Distrito Nacional (district) •capital of Dominican Republic •most populated city in Dominican Republic (2,138,000 people) •chief port on the Caribbean Sea

ENVIRONMENTAL/ECONOMIC:
Climate: tropical maritime with little temperature change
Natural Resources: nickel, bauxite, gold, silver
Agricultural Products: sugarcane, coffee, cotton, cacao, cattle
Major Exports: ferronickel, sugar, gold, silver, coffee
Natural Hazards: hurricanes, floods, droughts

EAST TIMOR
Country Name: Democratic Republic of Timor-Leste
Other Names: Timor-Leste
Continent: Asia
Area: 5,640 sq mi / 14,609 sq km
Population: 1,186,000
Population Density: 210 people per sq mi / 81 people per sq km
Capital: Dili

PHYSICAL:
Highest Point: Foho Tatamailau 9,721 ft / 2,963 m
Lowest Point: Timor Sea 0 ft / 0 m (sea level)
Peaks (minimum elevation of 8,000 ft / 2,400 m):
  Foho Tatamailau: •located in western East Timor •highest point in East Timor
Seas:
  Savu: •forms most of the northern coast of East Timor •feeds into the Indian Ocean
  Timor: •forms most of the southern coast of East Timor •feeds into the Indian Ocean

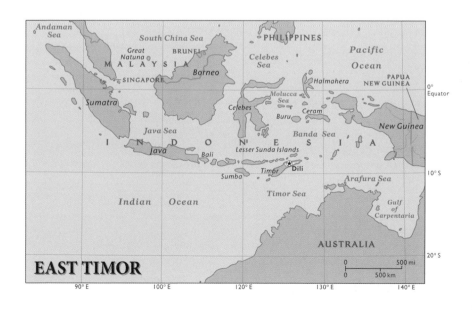

Straits:
    Wetar: •forms most of the northern coast of East Timor •separates East Timor from the island of Wetar

Islands:
    Timor: •main island of East Timor •bordered by the Savu Sea, Timor Sea, and Wetar Strait •western half belongs to Indonesia

POLITICAL:
Independence: May 20, 2002 (from Indonesia)
Former Name: Portuguese Timor
Bordering Country: Indonesia (1)
Administrative Divisions: Aileu, Ainaro, Baucau, Bobonaro (Maliana), Cova Lima (Suai), Dili, Ermera, Lautem (Los Palos), Liquica, Manatuto, Manufahi (Same), Oecussi-Ambeno, Viqueque (13 administrative districts)
Ethnic/Racial Groups: Austronesian, Papuan
Religion: Christianity (primarily Roman Catholic)
Languages: Tetum, Portuguese, Bahasa Indonesian, English
Currency: U.S. dollar
Current President: Jose Ramos-Horta
Cities:
    Dili: •located in northern East Timor •city in the Dili administrative district •capital of East Timor •most populated city in East Timor (166,000 people) •chief port on the Wetar Strait

ENVIRONMENTAL/ECONOMIC:
Climate: tropical, hot and humid with rainy/dry seasons
Natural Resources: gold, petroleum, natural gas, manganese, marble
Agricultural Products: coffee, rice, maize, cassava, sweet potatoes
Major Exports: coffee, sandalwood, marble
Natural Hazards: floods, landslides, earthquakes, tsunamis, tropical cyclones

ECUADOR
Country Name: Republic of Ecuador
Continent: South America
Area: 109,483 sq mi / 283,560 sq km
Population: 14,666,000
Population Density: 134 people per sq mi / 52 people per sq km (most dense in South America)
Capital: Quito

PHYSICAL:
Highest Point: Chimborazo 20,561 ft / 6,267 m
Lowest Point: Pacific Ocean 0 ft / 0 m (sea level)

Mountain Ranges:
  Andes: •cover much of Ecuador •extend into Colombia and Peru
Peaks (minimum elevation of 8,000 ft / 2,400 m):
  Chimborazo: •located in the Andes •highest point in Ecuador
Rain Forests:
  Amazon: •located in eastern Ecuador •extends into Colombia and Peru •largest rain forest in the world
Oceans:
  Pacific: •forms most of the western coast of Ecuador
Gulfs:
  Gulf of Guayaquil: •forms the southwestern coast of Ecuador •feeds into the Pacific Ocean
Rivers:
  Pastaza: •has its source in the Andes •extends into Peru
  Putumayo: •forms part of the Ecuador-Colombia border •extends into Colombia

POLITICAL:

Independence: May 24, 1822 (from Spain)
Bordering Countries: Peru, Colombia (2)
Administrative Divisions: Azuay, Bolivar, Canar, Carchi, Chimborazo, Cotopaxi, El Oro, Esmeraldas, Galapagos, Guayas, Imbabura, Loja, Los Rios, Manabi, Morona-Santiago, Napo, Orellana, Pastaza, Pichincha, Santa Elena, Santo Domingo de los Tsachilas, Sucumbios, Tungurahua, Zamora-Chinchipe (24 provinces)
Ethnic/Racial Groups: mestizo, Amerindian, black
Religion: Christianity (Roman Catholic)

Languages: Spanish, Quechua
Currency: U.S. dollar
Current President: Rafael Correa Delgado
Cities (capital, largest, or with at least a million people):
  Guayaquil: •located in western Ecuador •city in the Guayas province of Ecuador •most populated city in Ecuador (2,690,000 people) •chief port on the Gulf of Guayaquil
  Quito: •located in northern Ecuador •city in the Pichincha province of Ecuador •capital of Ecuador (1,801,000 people) •located in a valley in the Andes

ENVIRONMENTAL/ECONOMIC:
Climate: tropical along the coasts and in the rain forest; cooler in the highlands
Natural Resources: petroleum, fish, timber, hydropower
Agricultural Products: bananas, coffee, cacao, rice, cattle
Major Exports: petroleum, bananas, shrimp, coffee, cacao
Natural Hazards: earthquakes, landslides, volcanoes, floods, droughts

EGYPT
Country Name: Arab Republic of Egypt
Continent: Africa
Area: 386,874 sq mi / 1,002,000 sq km
Population: 82,637,000
Population Density: 214 people per sq mi / 82 people per sq km
Capital: Cairo

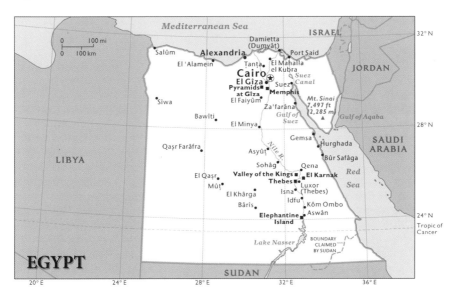

PHYSICAL:
Highest Point: Mount Catherine 8,625 ft / 2,629 m
Lowest Point: Qattara Depression 436 ft / 133 m below sea level
Peaks (minimum elevation of 8,000 ft / 2,400 m):
    Mount Catherine: •located on the Sinai Peninsula •highest point in Egypt
Plateau Regions:
    Libyan Plateau: •located in northwestern Egypt •extends into Libya
Depressions:
    Qattara Depression: •located in northwestern Egypt •lowest point in Egypt
Deserts:
    Eastern: •located in eastern Eqypt •extends into Sudan
    Libyan: •located in southwestern Egypt •extends into Libya and Sudan
    Sahara: •covers most of the country •extends into Libya and Sudan •largest desert in the world
    Western: •located in central Egypt •extends into Libya
Seas:
    Mediterranean: •forms most of the northern coast of Egypt •feeds into the Atlantic Ocean
    Red: •forms much of the eastern coast of Egypt •feeds into the Gulf of Aden
Gulfs:
    Gulf of Suez: •forms part of the northeastern coast of Egypt •feeds into the Red Sea
    Gulf of Aqaba: •forms part of the northeastern coast of Egypt •feeds into the Red Sea
Canals:
    Suez: •connects the Mediterranean and Red Seas •separates the Sinai Peninsula from the rest of Egypt
Lakes:
    Lake Nasser: •located in southeastern Egypt •on the border with Sudan •fed by the Nile River •formed by the Aswan High Dam
Rivers:
    Nile: •has its mouth in the Mediterranean Sea •extends into Sudan •longest river in the world
Deltas:
    Nile River Delta: •mouth of the Nile River •feeds into the Mediterranean Sea
Dams:
    Aswan High Dam: •located in southeastern Egypt •part of the Nile River •forms Lake Nasser
Peninsulas:
    Sinai: •bordered by the Red Sea, Gulf of Suez, Gulf of Aqaba, and Mediterranean Sea

POLITICAL:
Independence: February 28, 1922 (from the United Kingdom)
Former Name: United Arab Republic (with Syria)
Bordering Countries: Sudan, Libya, Israel (3)
Administrative Divisions: Ad Daqahliyah, Al Bahr al Ahmar, Al Buhayrah, Al Fayyum, Al Gharbiyah, Al Iskandariyah, Al Ismailiyah, Al Jizah, Al Minufiyah, Al Minya, Al Qahirah, Al Qalyubiyah, Al Uqsor, Al Wadi al Jadid, Ash Sharqiyah, As Suways, Aswan, Asyut, Bani Suwayf, Bur Said, Dumyat, Janub Sina, Kafr ash Shaykh, Matruh, Qina, Shamal Sina, Suhaj (27 governorates)
Ethnic/Racial Groups: Egyptian Arab
Religions: Islam (primarily Sunni), Christianity (primarily Coptic)
Languages: Arabic, English, French
Currency: Egyptian pound
Current Prime Minister: Essam Abdel Aziz Sharaf
Cities (capital, largest, or with at least a million people):
 Cairo: •located in northern Egypt •city in the Al Qahirah governorate •capital of Egypt •most populated city in Egypt (10,903,000 people) •chief port on the Nile River
 Alexandria: •located in northern Egypt •city in the Al Iskandariyah governorate •chief port on the Mediterranean Sea
 Al Jizah (Giza): •located in northern Egypt •city in the Al Jizah governorate
 Shubra al Khaymah: •located in northern Egypt •city in the Al Qalyubiyah governorate

ENVIRONMENTAL/ECONOMIC:
Climate: desert; hot and dry summers/moderate winters
Natural Resources: petroleum, natural gas, iron ore, phosphates, manganese
Agricultural Products: cotton, rice, corn, wheat, cattle
Major Exports: crude oil and petroleum products, cotton, textiles, metal products, chemicals
Natural Hazards: droughts, earthquakes, flash floods, landslides, khamsin (windstorms), dust storms, sandstorms

EL SALVADOR
Country Name: Republic of El Salvador
Continent: North America
Area: 8,124 sq mi / 21,041 sq km
Population: 6,227,000
Population Density: 766 people per sq mi / 296 people per sq km
Capital: San Salvador

PHYSICAL:
Highest Point: Cerro El Pital 8,957 ft / 2,730 m
Lowest Point: Pacific Ocean 0 ft / 0 m (sea level)
Peaks (minimum elevation of 8,000 ft / 2,400 m):

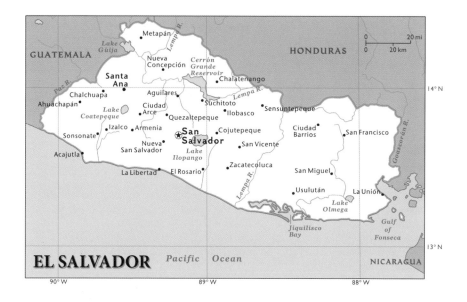

Cerro El Pital: •located in northwestern El Salvador •on the El Salvador-Honduras border •highest point in El Salvador

Santa Ana Volcano: •located in western El Salvador

Oceans:
Pacific: •forms most of the southern coast of El Salvador

Gulfs:
Gulf of Fonseca: •forms part of the southeastern coast of El Salvador •feeds into the Pacific Ocean

POLITICAL:

Independence: September 15, 1821 (from Spain)
Bordering Countries: Honduras, Guatemala (2)
Administrative Divisions: Ahuachapan, Cabanas, Chalatenango, Cuscatlan, La Libertad, La Paz, La Union, Morazan, San Miguel, San Salvador, Santa Ana, San Vicente, Sonsonate, Usulutan (14 departments)
Ethnic/Racial Groups: mestizo, white, Amerindian
Religion: Christianity (Roman Catholic, Evangelical)
Languages: Spanish, Nahua
Currency: U.S. dollar
Current President: Carlos Mauricio Funes Cartagena
Cities (capital, largest, or with at least a million people):
San Salvador: •located in central El Salvador •city in the San Salvador department •capital of El Salvador •most populated city in El Salvador (1,534,000 people)

ENVIRONMENTAL/ECONOMIC:
Climate: tropical; temperate in the uplands with rainy/dry seasons
Natural Resources: hydropower, geothermal power, petroleum, arable land
Agricultural Products: coffee, sugar, corn, rice, shrimp
Major Exports: offshore-assembly exports, coffee, sugar, shrimp, textiles
Natural Hazards: earthquakes, volcanoes, hurricanes

EQUATORIAL GUINEA
Country Name: Republic of Equatorial Guinea
Continent: Africa
Area: 10,831 sq mi / 28,051 sq km
Population: 720,000
Population Density: 66 people per sq mi / 26 people per sq km
Capital: Malabo

PHYSICAL:
Highest Point: Pico Basile (Pico de Santa Isabel) 9,869 ft / 3,008 m
Lowest Point: Gulf of Guinea 0 ft / 0 m (sea level)
Mountain Ranges:
 Monts de Cristal: •located in southern Equatorial Guinea •extend into Gabon
Peaks (minimum elevation of 8,000 ft / 2,400 m):
 Pico de Santa Isabel: •located on the island of Bioko •highest point in Equatorial Guinea
Gulfs:
 Gulf of Guinea: •forms most of the western coast of Equatorial Guinea •feeds into the Atlantic Ocean
Bays:
 Bight of Bonny: •forms part of the western coast of Equatorial Guinea •feeds into the Gulf of Guinea
Islands:
 Bioko: •an island off the coast of Cameroon •bordered by the Gulf of Guinea and the Bight of Bonny

POLITICAL:
Independence: October 12, 1968 (from Spain)
Former Name: Spanish Guinea
Bordering Countries: Gabon, Cameroon (2)
Regions:
 Rio Muni: •includes all of mainland Equatorial Guinea
Administrative Divisions: Annobon, Bioko Norte, Bioko Sur, Centro Sur, Kie-Ntem, Litoral, Wele-Nzas (7 provinces)
Ethnic/Racial Groups: Fang, Bubi
Religions: Christianity (predominantly Roman Catholic), pagan practices
Languages: Spanish, French, pidgin English, Fang, Bubi, Ibo

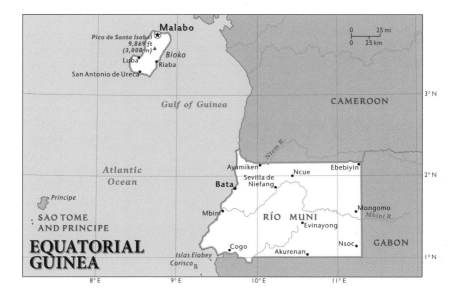

Currency: Communaute Financiere Africaine franc
Current President: Teodoro Obiang Nguema Mbasogo
Cities (capital, largest, or with at least a million people):
  Malabo: •located on the island of Bioko •city in the Bioko Norte province •capital of Equatorial Guinea •most populated city in Equatorial Guinea (128,000 people) •chief port on the Bight of Bonny

ENVIRONMENTAL/ECONOMIC:
Climate: tropical, hot and humid
Natural Resources: oil, petroleum, timber, gold, manganese
Agricultural Products: coffee, cocoa, rice, yarns, livestock
Major Exports: petroleum, methanol, timber, cacao
Natural Hazards: windstorms, flash floods

ERITREA
Country Name: State of Eritrea
Continent: Africa
Area: 46,774 sq mi / 121,144 sq km
Population: 5,939,000
Population Density: 127 people per sq mi / 49 people per sq km
Capital: Asmara

PHYSICAL:
Highest Point: Soira 9,902 ft / 3,018 m
Lowest Point: Denakil Depression 381 ft / 116 m below sea level
Mountain Ranges:
  Ethiopian Highlands: •stretch across western and central Eritrea •extend into Ethiopia
Peaks (minimum elevation of 8,000 ft / 2,400 m):
  Soira: •located in the Ethiopian Highlands •highest point in Eritrea
Depressions:
  Denakil: •extends into Ethiopia •lowest point in Eritrea
Seas:
  Red: •forms most of the coast of Eritrea •feeds into the Gulf of Aden
Straits:
  Bab al Mandab: •forms part of the southeastern coast of Eritrea •separates Eritrea from Yemen •connects the Red Sea and the Gulf of Aden
Regions:
  Denakil: •lowland, geothermal region in central and eastern Eritrea •extends into Ethiopia and Djibouti •located to the east of the Ethiopian Highlands

POLITICAL:
Independence: May 24, 1993 (from Ethiopia)
Former Name: Eritrea Autonomous Region in Ethiopia
Bordering Countries: Ethiopia, Sudan, Djibouti (3)
Administrative Divisions: Anseba, Debub, Debubawi K'eyih Bahri, Gash Barka, Ma'akel, Semenawi Keyih Bahri (6 regions)

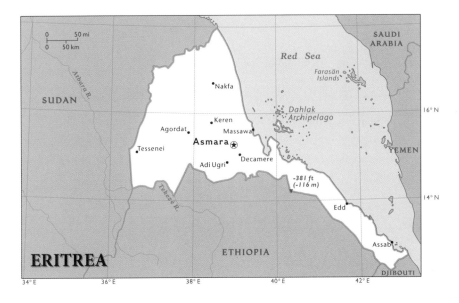

COUNTRIES A TO Z | PAGE 119

Ethnic/Racial Groups: Tigrinya, Tigre, Kunama, Afar, Saho
Religions: Islam, Christianity (Coptic, Roman Catholic, Protestant)
Languages: Afar, Arabic, Tigre, Kunama, Tigrinya
Currency: nafka
Current President: Isaias Afwerki
Cities (capital, largest, or with at least a million people):
  Asmara: •located in central Eritrea •city in the Central region •capital of Eritrea •most populated city in Eritrea (649,000 people)

ENVIRONMENTAL/ECONOMIC:
Climate: arid to semiarid; hot and dry in the desert along the Red Sea; cooler and wetter in the highlands
Natural Resources: gold, potash, zinc, copper, salt
Agricultural Products: sorghum, lentils, vegetables, corn, livestock
Major Exports: livestock, sorghum, textiles, food, small manufactures
Natural Hazards: droughts, locust swarms

ESTONIA
Country Name: Republic of Estonia
Continent: Europe
Area: 17,462 sq mi / 45,227 sq km
Population: 1,340,000
Population Density: 77 people per sq mi / 30 people per sq km
Capital: Tallinn

PHYSICAL:
Highest Point: Suur Munamagi 1,043 ft / 318 m
Lowest Point: Baltic Sea 0 ft / 0 m (sea level)
Seas:
  Baltic: •forms part of the western coast of Estonia •feeds into the Kattegat
Gulfs:
  Gulf of Finland: •forms the northern coast of Estonia •feeds into the Baltic Sea
  Gulf of Riga: •forms the southwestern coast of Estonia •forms part of the western coast of Estonia •feeds into the Baltic Sea
Lakes:
  Lake Peipus: •located in eastern Estonia •on the Estonia-Russia border
  Lake Pskov: •southern part of Lake Peipus •on the Estonia-Russia border
  Vorts: •located in central Estonia
Islands:
  Hiiumaa: •located off the western coast of Estonia •bordered by the Baltic Sea
  Saaremaa: •located off the southwestern coast of Estonia •bordered by the Baltic Sea and Gulf of Riga

POLITICAL:
Independence: August 20, 1991 (from Soviet Union)
Former Name: Estonian Soviet Socialist Republic
Bordering Countries: Latvia, Russia (2)
Administrative Divisions: Harjumaa, Hiiumaa, Ida-Virumaa, Jarvamaa, Jogevamaa, Laanemaa, Laane-Virumaa, Parnumaa, Polvamaa, Raplamaa, Saaremaa, Tartumaa, Valgamaa, Viljandimaa, Vorumaa (15 counties)
Ethnic/Racial Groups: Estonian, Russian
Religion: Christianity (Evangelical Lutheran, Russian Orthodox, Eastern Orthodox)
Languages: Estonian, Russian, Ukrainian
Currency: euro
Current President: Toomas Hendrik Ilves
Cities (capital, largest, or with at least a million people):
  Tallinn: •located in northern Estonia •city in the Harjumaa county •capital of Estonia •most populated city in Estonia (399,000 people) •chief port on the Gulf of Finland

ENVIRONMENTAL/ECONOMIC:
Climate: maritime, wet and moderate winters /cool summers
Natural Resources: oil shale, peat, phosphorite, clay, limestone
Agricultural Products: potatoes, vegetables, livestock and dairy products, fish
Major Exports: machinery and equipment, wood and paper, textiles, food products, furniture
Natural Hazards: floods

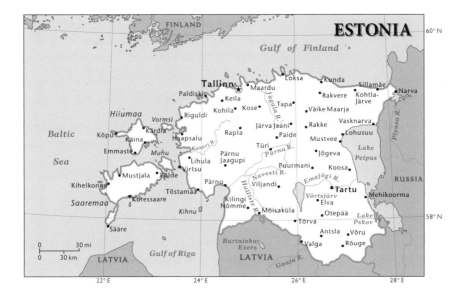

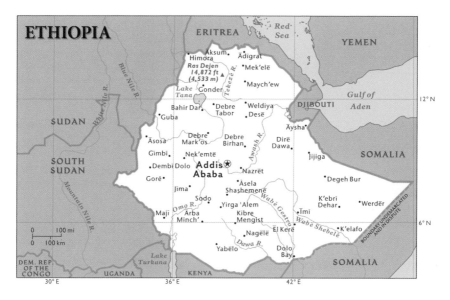

ETHIOPIA

Country Name: Federal Democratic Republic of Ethiopia
Continent: Africa
Area: 437,600 sq. mi / 1,133,380 sq. km
Population: 87,118,000
Population Density: 199 people per sq mi / 77 people per sq km
Capital: Addis Ababa

PHYSICAL:
Highest Point: Ras Dejen 14,872 ft / 4,533 m
Lowest Point: Denakil Depression 410 ft / 125 m below sea level
Mountain Ranges:
 Ethiopian Highlands: •cover most of western and central Ethiopia •extend into Eritrea
Peaks (minimum elevation of 8,000 ft / 2,400 m):
 Ras Dejen: •located in the Ethiopian Highlands •highest point in Ethiopia
Valleys:
 Great Rift: •covers an area from northeastern to southwestern Ethiopia •surrounded by the Ethiopian Highlands •extends from Ethiopia into Djibouti and Kenya
Depressions:
 Denakil Depression: •located in northern Ethiopia •extends into Eritrea •contains the lowest point in Ethiopia •location of Dalol, the world's hottest place

Lakes:
    Lake Tana: •located in northwestern Ethiopia •located high in the Ethiopian Highlands •source of the Blue Nile
    Lake Turkana: •located in southwestern Ethiopia •lake in the Great Rift Valley on the edge of the Ethiopian Highlands •fed by the Omo River •extends into Kenya

Rivers:
    Awash: •has its mouth in Lake Abbe •in the Great Rift Valley •has its source in the Ethiopian Highlands
    Blue Nile: •has its source in Lake Tana •in the Ethiopian Highlands •extends into Sudan
    Omo: •has its mouth in Lake Turkana •has its source in the Ethiopian Highlands
    Wabe Shebele: •has its source in the Ethiopian Highlands •extends into Somalia

Regions:
    Denakil: •lowland, geothermal region in northeastern Africa •east of the Ethiopian Highlands •extends into Eritrea and Djibouti •location of Dalol, the world's hottest place
    Haud: •lowland, desert region in eastern Ethiopia •east of the Ethiopian Highlands •extends into Somalia
    Ogaden: •lowland, desert region in southeastern Ethiopia •east of the Ethiopian Highlands

POLITICAL:

Independence: the oldest independent country in Africa (since the early 1800s) and one of the oldest in the world
Former Names: Abyssinia, Italian East Africa
Bordering Countries: Sudan, South Sudan, Somalia, Eritrea, Kenya, Djibouti (6)
Regions:
    Afar: •region in northeastern Ethiopia •located in the lowland region of Denakil
    Amhara: •region in western and central Ethiopia •located in the Ethiopian Highlands
    Boran: •region in southern Ethiopia •located to the south of the Ethiopian Highlands •extends into Kenya
    Oromia: •region in central Ethiopia •located in the Ethiopian Highlands and the Great Rift Valley
    Somali: •region in eastern Ethiopia •located in the lowland regions of Haud and Ogaden •located east of the Ethiopian Highlands
    Tigray: •region in northern Ethiopia •located in the Ethiopian Highlands and the Denakil region

Administrative Divisions: Afar, Amara, Binshangul Gumuz, Gambela Hizboch, Hareri Hizb, Oromiya, Sumale, Tigray, Yedebub (9 ethnically based states) •Addis Ababa, Dire Dawa (2 self-governing administrations)
Ethnic/Racial Groups: Oromo, Amhara, Tigre, Sidamo, Shankella, Somali, Afar, Gurage
Religions: Islam, Christianity (Ethiopian Orthodox), animism
Languages: Amharic, Tigrinya, Orominga, Guaraginga, Somali, Arabic
Currency: birr

Current President: Girma Woldegiorgis
Cities (capital, largest, or with at least a million people):
  Addis Ababa: •located in central Ethiopia •city in the Addis Ababa self-governing administration •capital of Ethiopia •most populated city in Ethiopia (2,863,000 people) •in the Ethiopian Highlands

ENVIRONMENTAL/ECONOMIC:
Climate: tropical monsoon; variance in climates
Natural Resources: gold, platinum, copper, potash, natural gas
Agricultural Products: cereals, pulses, coffee, oilseed, cattle
Major Exports: coffee, qat, gold, leather products, live animals
Natural Hazards: earthquakes, volcanoes, droughts

FIJI

Country Name: Republic of the Fiji Islands
Continent: Australia/Oceania
Area: 7,095 sq mi / 18,376 sq km
Population: 852,000
Population Density: 120 people per sq mi / 46 people per sq km
Capital: Suva

PHYSICAL:

Highest Point: Tomanivi 4,341 ft / 1,323 m
Lowest Point: Pacific Ocean 0 ft / 0 m (sea level)
Oceans:
 Pacific: •forms most of the coast of the Fiji Islands
Islands:
 Vanua Levu: •second largest island in the Fiji Islands •bordered by the Pacific Ocean
 Viti Levu: •largest island in the Fiji Islands •bordered by the Pacific Ocean

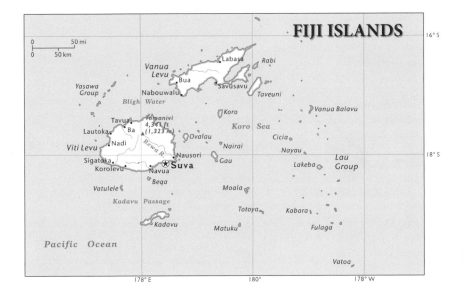

POLITICAL:
Independence: October 10, 1970 (from the United Kingdom)
Administrative Divisions: Central, Eastern, Northern, Western (4 divisions)
 • Rotuma (1 dependency)
Ethnic/Racial Groups: Fijian, Indian
Religions: Christianity, Hinduism, Islam
Languages: English, Fijian, Hindustani
Currency: Fijian dollar
Current President: Ratu Epeli Nailatikau
Cities: (capital, largest, or with at least a million people):
 Suva: • located on the island of Viti Levu • city in the Central division • capital of the Fiji Islands • most populated city in the Fiji Islands (174,000 people) • chief port on the Pacific Ocean

ENVIRONMENTAL/ECONOMIC:
Climate: tropical marine, little temperature variation
Natural Resources: timber, fish, gold, copper, offshore oil potential
Agricultural Products: sugarcane, coconuts, cassava (tapioca), rice
Major Exports: sugar, garments, gold, timber, fish
Natural Hazards: cyclones

FINLAND
Country Name: Republic of Finland
Continent: Europe
Area: 130,558 sq mi / 338,145 sq km
Population: 5,387,000
Population Density: 41 people per sq mi / 16 people per sq km
Capital: Helsinki

PHYSICAL:
Highest Point: Halti 4,357 ft / 1,328 m
Lowest Point: Baltic Sea 0 ft / 0 m (sea level)
Mountain Ranges:
 Kjolen: • located in northwestern Finland • extend into Norway and Sweden
Plateau Regions:
 Maanselka: • covers an area from northeastern to southeastern Finland • extends into Russia and Norway
Seas:
 Baltic: • forms part of the southwestern coast of Finland • feeds into the Kattegat
Gulfs:
 Gulf of Bothnia: • forms most of the western coast of Finland • feeds into the Baltic Sea
 Gulf of Finland: • forms most of the southern coast of Finland • feeds into the Baltic Sea

Islands:
  Aland •located off the coast of southwestern Finland •bordered by the Baltic Sea and Gulf of Bothnia

POLITICAL:
Independence: December 6, 1917 (from Russia)
Bordering Countries: Russia, Norway, Sweden (3)
Regions:
  Lapland: •region in northern Finland •southeast of Kjolen •extends into Norway, Sweden, and Russia
Administrative Divisions: Ahvenanmaa, Etela-Karjala, Etela-Pohjanmaa, Etela-Savo, Kanta-Hame, Kainuu, Keski-Pohjanmaa, Keski-Suomi, Kymenlaakso, Lappi, Paijat-Hame, Pirkanmaa, Pohjanmaa, Pohjois-Karjala, Pohjois-Pohjanmaa, Pohjois-Savo, Satakunta, Uusimaa, Varsinais-Suomi (19 regions)
Ethnic/Racial Groups: Finnish, Swedish
Religion: Christianity (primarily Evangelical Lutheran)
Languages: Finnish, Swedish
Currency: euro
Current President: Tarja Halonen
Cities (capital, largest, or with at least a million people):
  Helsinki: •located in southern Finland •city in the Etela-Suomen Laani province •capital of Finland •most populated city in Finland (1,107,000 people) •chief port on the Gulf of Finland

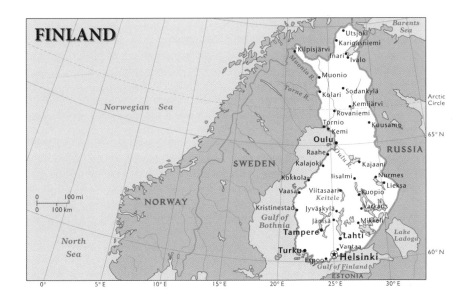

ENVIRONMENTAL/ECONOMIC:
Climate: cold temperate, potentially subarctic, moderated by the North Atlantic Current, the Baltic Sea, and many lakes
Natural Resources: timber, iron ore, copper, lead, zinc
Agricultural Products: barley, wheat, sugar beets, potatoes, dairy cattle
Major Exports: machinery and equipment, chemicals, metals, timber, paper
Natural Hazards: snowstorms

FRANCE
Country Name: French Republic
Continent: Europe
Area: 210,026 sq mi / 543,965 sq km
Population: 63,305,000
Population Density: 301 people per sq mi / 116 people per sq km
Capital: Paris

PHYSICAL:
Highest Point: Mont Blanc 15,781 ft / 4,810 m
Lowest Point: Rhone River Delta 7 ft / 2 m below sea level
Mountain Ranges:
  Alps: •located in southeastern France •extend into Switzerland and Italy
  Cevennes: •located in southern France
  Jura: •located in eastern France •extend into Switzerland
  Massif Central: •located in southern France
  Pyrenees: •located in southwestern France •extend into Spain
  Vosges: •located in northeastern France
Peaks (minimum elevation of 8,000 ft / 2,400 m):
  Mont Blanc: •located in the Alps •highest point in France
Plateau Regions:
  Plateau de Langres: •located in eastern France
Oceans:
  Atlantic: •forms part of the western coast of France
Seas:
  Ligurian: •forms part of the southeastern coast of France •feeds into the Mediterranean Sea
  Mediterranean: •forms part of the southeastern coast of France •feeds into the Atlantic Ocean
  North: •forms part of the northern coast of France •feeds into the Atlantic Ocean
  Tyrrhenian: •forms most of the eastern coast of the island of Corsica
Gulfs:
  Gulf of Lion: •forms part of the southern coast of France • feeds into the Mediterranean Sea
  Gulf of Saint-Malo: •forms part of the northwestern coast of France •feeds into the English Channel

Bays:
    Bay of Biscay: • forms part of the western coast of France • feeds into the Atlantic Ocean
    Bay of the Seine: • forms part of the northwestern coast of France • feeds into the English Channel

Straits:
    Strait of Bonifacio: • forms part of the southern coast of the island of Corsica • separates Corsica from the island of Sardinia (Italy) • connects the Mediterranean and Tyrrhenian Seas
    Strait of Dover: • forms part of the northern coast of France • separates France from the United Kingdom • connects the North Sea and the English Channel

Channels:
    English: • forms most of the northern coast of France • separates France from the United Kingdom • connects the North Sea and the Atlantic Ocean

Lakes:
    Lake Geneva: • located in eastern France • on the France-Switzerland border • surrounded by the Jura Mountains and the Alps

Rivers:
    Allier: • tributary of the Loire River • has its source in Massif Central
    Durance: • tributary of the Rhone River • has its source in the Alps
    Garonne: • has its mouth in the Gironde estuary • has its source in the Pyrenees
    Loire: • has its mouth in the Bay of Biscay • has its source in Massif Central
    Meuse: • has its source in northeastern France • extends into Belgium
    Moselle: • has its source in the Vosges • extends into Germany

Rhine: •forms part of the France-Germany border •extends into Germany and Switzerland
Rhone: •has its mouth in the Gulf of Lion •extends into Switzerland
Saone: •tributary of the Rhone River •has its source in northeastern France
Seine: •has its mouth in the Bay of the Seine •has its source on the Plateau de Langres
Deltas:
Gironde Estuary: •mouth of the Garonne River •feeds into the Bay of Biscay
Rhone River Delta: •mouth of the Rhone River •feeds into the Gulf of Lion
Peninsulas:
Cotentin: •bordered by the English Channel, Gulf of Saint-Malo, and Bay of the Seine
Islands:
Corsica: •in the Mediterranean Sea •off the southeastern coast of France •region of France •separated from the Italian island of Sardinia by the Strait of Bonifacio
Regions:
Landes: •lowland region in southwestern France •east of the Bay of Biscay

POLITICAL:
Independence: 486 (unified by Clovis)
Bordering Countries: Spain, Belgium, Switzerland, Italy, Germany, Luxembourg, Andorra, Monaco (8)
Regions:
Alsace: •region in northeastern France •east of the Vosges •west of the Rhine River •north of the Jura Mountains
Angoumois: •region in western France
Anjou: •region in western France
Ardennes: •region in northeastern France •extends into Belgium
Artois: •region in northern France •east of the English Channel
Auvergne: •region in central France •located in the Massif Central
Bearn: •region in southwestern France •southern half is in the Pyrenees
Berry: •region in central France
Bourbonnais: •region in central France •north of the Massif Central
Brittany: •region in northwestern France •south of the English Channel •east of the Atlantic Ocean •north of the Bay of Biscay
Burgundy: •region in eastern France •south of the Plateau de Langres •northeast of the Massif Central •west of the Jura Mountains and the Alps
Champagne: •region in northern France
Dauphine: •region in southeastern France •located in the Alps
Flanders: •region in northern France •south of the North Sea •extends into Belgium
Foix: •region in southern France •north of the Pyrenees
Franche: •region in eastern France •west of the Jura Mountains
Gascony: •region in southwestern France •east of the Bay of Biscay •north of the Pyrenees

Guienne: •region that stretches across southern France •east of the Bay of Biscay •eastern part is in the Massif Central
Ile-de-France: •region in northern France •Paris is located here
Languedoc: •region in southern France •includes the Cevennes •southeast of the Massif Central
Limousin: •region in central France •west of the Massif Central
Lorraine: •region in northeastern France •west of the Vosges
Lyonnais: •region in eastern France •northeast of the Massif Central
Maine: •region in northwestern France
Marche: •region in central France •north of the Massif Central
Nivernais: •region in central France
Normandy: •region in northwestern France •south of the English Channel •east of the Gulf of Saint-Malo •includes the Cotentin Peninsula
Orleanais: •region in central France
Perche: •region in northern France
Picardy: •region in northern France •east of the English Channel
Poitou: •region in western France •east of the Bay of Biscay
Provence: •region in southeastern France •eastern part is located in the Alps •north of the Mediterranean Sea •northeast of the Gulf of Lion •northwest of the Ligurian Sea
Roussillon: •region in southern France •western part is in the Pyrenees •west of the Mediterranean Sea
Saintonge: •region in western France •east of the Bay of Biscay
Savoy: •region in eastern France •located in the Alps

Administrative Divisions: Alsace, Aquitaine, Auvergne, Basse-Normandie, Bourgogne, Bretagne, Centre, Champagne-Ardenne, Corse (Corsica), Franche-Comte, Haute-Normandie, Ile-de-France, Languedoc-Roussillon, Limousin, Lorraine, Midi-Pyrenees, Nord-Pas-de-Calais, Pays de la Loire, Picardie, Poitou-Charentes, Provence-Alpes-Cote d'Azur, Rhone-Alpes (22 regions)

External Territories: French Guiana, French Polynesia, French Southern and Antarctic Territories, Guadeloupe, Martinique, Mayotte, New Caledonia, Reunion, Saint-Pierre and Miquelon, Wallis and Futuna (10)

Ethnic/Racial Groups: French, with Slavic, North African, Indochinese, and Basque minorities

Religion: Christianity (primarily Roman Catholic)

Language: French

Currency: euro

Current President: Nicolas Sarkozy

Cities (capital, largest, or with at least a million people):
Paris: •located in northern France •city in the Ile-de-France region •capital of France •most populated city in France (10,410,000 people) •chief port on the Seine River
Lyon •located in eastern France •city in the Rhone-Alpes region •located at the junction of the Rhone and Saone Rivers

Marseille •located in southeastern France •city in the Provence-Alpes-Cote d'Azur region •chief port on the Gulf of Lion

ENVIRONMENTAL/ECONOMIC:
Climate: Cool winters/mild summers with mild winters/hot summers along the Mediterranean; occasional mistral winds
Natural Resources: coal, iron ore, bauxite, zinc, uranium
Agricultural Products: wheat, cereals, sugar beets, potatoes, beef
Major Exports: machinery and transportation equipment, aircraft, plastics, chemicals, pharmaceuticals
Natural Hazards: floods, avalanches, windstorms, droughts, forest fires

GABON

Country Name: Gabonese Republic
Continent: Africa
Area: 103,347 sq mi / 267,667 sq km
Population: 1,534,000
Population Density: 15 people per sq mi / 6 people per sq km
Capital: Libreville

PHYSICAL:
Highest Point: Mount Iboundji 5,167 ft / 1,575 m
Lowest Point: Atlantic Ocean 0 ft / 0 m (sea level)
Mountain Ranges:
   Monts de Cristal: •cover much of Gabon •extend into Equatorial Guinea and Republic of the Congo
Oceans:
   Atlantic: •forms part of the western coast of Gabon
Gulfs:
   Gulf of Guinea: •forms part of the western coast of Gabon •feeds into the Atlantic Ocean

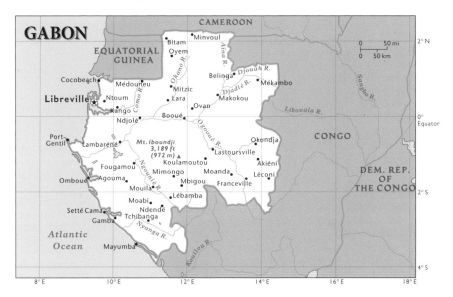

POLITICAL:
Independence: August 17, 1960 (from France)
Bordering Countries: Republic of the Congo, Equatorial Guinea, Cameroon (3)
Administrative Divisions: Estuaire, Haut-Ogooue, Moyen-Ogooue, Ngounie, Nyanga, Ogooue-Ivindo, Ogooue-Lolo, Ogooue-Maritime, Woleu-Ntem (9 provinces)
Ethnic/Racial Groups: Fang, Bapounou, Nzebi, Obama, European
Religions: Christianity, indigenous beliefs
Languages: French, Fang, Myene, Nzebi, Bapounou/Eschira, Bandjabi
Currency: Communaute Financiere Africaine franc
Current President: Ali Ben Bongo Ondimba
Cities (capital, largest, or with at least a million people):
  Libreville: • located in northwestern Gabon • city in the Estuaire province • capital of Gabon • most populated city in Gabon (619,000 people) • chief port on the Gulf of Guinea

ENVIRONMENTAL/ECONOMIC:
Climate: tropical, hot and humid
Natural Resources: petroleum, manganese, uranium, gold, timber
Agricultural Products: cacao, coffee, sugar, palm oil, cattle
Major Exports: crude oil, timber, manganese, uranium
Natural Hazards: floods, thunderstorms

GAMBIA
Country Name: Republic of the Gambia
Continent: Africa
Area: 4,361 sq mi / 11,295 sq km
Population: 1,778,000
Population Density: 408 people per sq mi / 157 people per sq km
Capital: Banjul

PHYSICAL:
Highest Point: unnamed location 174 ft / 53 m
Lowest Point: Atlantic Ocean 0 ft / 0 m (sea level)
Oceans:
  Atlantic: • forms the western coast of the Gambia
Rivers:
  Gambia: • has its mouth in the Atlantic Ocean • extends into Senegal

POLITICAL:
Independence: February 18, 1965 (from the United Kingdom)
Bordering Country: Senegal (1)

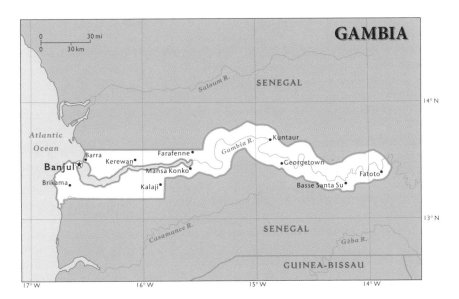

Administrative Divisions: Central River, Lower River, North Bank, Upper River, Western (5 divisions) • Banjul (1 city)
Ethnic/Racial Groups: Mandinka, Fula, Wolof, Jola, Serahuli
Religions: Islam, Christianity
Languages: English, Mandinka, Wolof, Fula
Currency: dalasi
Current President: Yahya A.J.J. Jammeh
Cities (capital, largest, or with at least a million people):
  Banjul: • located in western Gambia • city in the Banjul city division • capital of Gambia • most populated city in Gambia (436,000 people) • chief port on the Gambia River and the Gulf of Guinea

ENVIRONMENTAL/ECONOMIC:
Climate: tropical with rainy/dry seasons
Natural Resources: fish
Agricultural Products: rice, millet, sorghum, peanuts, cattle
Major Exports: peanut products, fish, palm kernels, re-exports
Natural Hazards: droughts

GEORGIA
Country Name: Republic of Georgia
Continent: Asia

Area: 26,911 sq mi / 69,700 sq km
Population: 4,329,000
Population Density: 161 people per sq mi / 62 people per sq km
Capital: Tbilisi

PHYSICAL:
Highest Point: Mta Shkhara 17,064 ft / 5,201 m
Lowest Point: Black Sea 0 ft / 0 m (sea level)
Mountain Ranges:
 Caucasus: •cover much of northern Georgia •extend into Russia and Azerbaijan
 Lesser Caucasus: •cover much of southern Georgia •extend into Turkey, Armenia, and Azerbaijan
Peaks (minimum elevation 8,000 ft / 2,400 m):
 Mta Shkhara: •located in the Caucasus •on the Georgia-Russia border •highest point in Georgia
Seas:
 Black: •forms the western coast of Georgia •feeds into the Mediterranean Sea

POLITICAL:
Independence: April 9, 1991 (from Soviet Union)
Former Name: Georgian Soviet Socialist Republic
Bordering Countries: Russia, Azerbaijan, Turkey, Armenia (4)
Administrative Divisions: Guria, Imereti, Kakheti, Kvemo Kartli, Mtskheta-Mtianeti, Racha-Lechkhumi and Kvemo Svaneti, Samegrelo-Zemo Svaneti,

Samtskhe-Javakheti, Shida Kartli (9 regions) •Tbilisi (1 city) •Abkhazia, Ajaria (2 autonomous republics)
Ethnic/Racial Groups: Georgian, Armenian, Russian, Azeri
Religions: Christianity (Georgian Orthodox, Russian Orthodox), Islam
Languages: Georgian, Russian, Armenian, Azeri
Currency: lari
Current President: Mikheil Saakashvili
Cities (capital, largest, or with at least a million people):
  Tbilisi: •located in southeastern Georgia •city in the Tbilisi city division •capital of Georgia •most populated city in Georgia (1,115,000 people) •between the Caucasus and Lesser Caucasus

ENVIRONMENTAL/ECONOMIC:
Climate: mild; Mediterranean-like along the Black Sea coast
Natural Resources: forests, hydropower, manganese deposits, iron ore, copper
Agricultural Products: citrus, grapes, tea, hazelnuts, livestock
Major Exports: scrap metal, machinery, chemicals, fuel re-exports, citrus fruits, tea
Natural Hazards: earthquakes

GERMANY
Country Name: Federal Republic of Germany
Continent: Europe
Area: 137,847 sq mi / 357,022 sq km
Population: 81,755,000 (largest of any country entirely in Europe)
Population Density: 593 people per sq mi / 229 people per sq km
Capital: Berlin

PHYSICAL:
Highest Point: Zugspitze 9,721 ft / 2,963 m
Lowest Point: Neuendorf bei Wilster 12 ft / 3.54 m below sea level
Mountain Ranges:
  Alps: •located in southern Germany •extend into Austria and Switzerland
  Erzgebirge: •located in eastern Germany •extend into the Czech Republic
  Harz: •located in central Germany
  Schwabische Alb: •located in southern Germany
  Schwarzwald: •located in southwestern Germany
  Thuringer Wald: •located in central Germany
Peaks (minimum elevation 8,000 ft / 2,400 m):
  Zugspitze: •located in the Alps •on the Germany-Austria border
Seas:
  Baltic: •forms part of the northeastern coast of Germany •feeds into the Kattegat
  North: •forms part of the northwestern coast of Germany •feeds into the Atlantic Ocean

Bays:
    Helgolander: • forms part of the northwestern coast of Germany • feeds into the North Sea
    Kiel: • forms part of the northern coast of Germany • feeds into Mecklenburger Bay
    Mecklenburger: • forms part of the northern coast of Germany • feeds into the Baltic Sea
    Pomeranian : • forms part of the northeastern coast of Germany • feeds into the Baltic Sea

Lakes:
    Lake Constance: • located in southern Germany • on the border with Switzerland and Austria • fed by the Rhine River

Rivers:
    Danube: • has its source in the Schwabische Alb • forms part of the Germany-Austria border • extends into Austria
    Elbe: • has its mouth in Helgolander Bay • extends into the Czech Republic
    Main: • tributary of the Rhine River • has its source in southern Germany
    Mosel: • tributary of the Rhine River • forms part of the Germany-Luxembourg border • extends into France
    Oder: • forms part of the Germany-Poland border • extends into Poland
    Rhine: • forms part of Germany's border with France and Switzerland • extends into the Netherlands and Switzerland
    Ruhr: • tributary of the Rhine River • has its source in western Germany
    Weser: • has its mouth in Helgolander Bay • has its source in central Germany

Lagoons:
    Stettiner Haff: •feeds into the Baltic Sea •extends into Poland
Islands:
    East Frisian: •island group off the northwestern coast of Germany •bordered by the North Sea and Helgolander Bay
    North Frisian: •island group off the northwestern coast of Germany •bordered by the North Sea and Helgolander Bay •northern islands in this group belong to Denmark
    Rugen: •island off the northeastern coast of Germany •bordered by the Baltic Sea
Regions:
    Eifel: •mountainous region in western Germany
    Black Forest: •mountainous region in southwestern Germany

POLITICAL:

Independence: January 18, 1871 (German Empire unification)
Former Names: German Empire, German Republic, German Reich
Bordering Countries: Austria, Czech Republic, Netherlands, Poland, France, Switzerland, Belgium, Luxembourg, Denmark (9)
Regions:
    Franconia: •region in southern Germany •north of the Schwabische Alb
    Friesland: •region in northwestern Germany •south of the North Sea and Helgolander Bay •extends into the Netherlands
    Pomerania: •region in northeastern Germany •south of the Baltic Sea and Pomeranian Bay •extends into Poland
Administrative Divisions: Baden-Wurttemberg, Bavaria, Berlin, Brandenburg, Bremen, Hamburg, Hessen, Lower Saxony, Mecklenburg-Western Pomerania, North Rhine-Westphalia, Rhineland-Palatinate, Saarland, Saxony, Saxony-Anhalt, Schleswig-Holstein, Thuringia (16 states)
Ethnic/Racial Groups: German, Turkish
Religion: Christianity (Protestant, Roman Catholic)
Language: German
Currency: euro
Current President: Christian Wulff; **Chancellor:** Angela Merkel
Cities (capital, largest, or with at least a million people):
    Berlin: •located in northeastern Germany •city in the Berlin state •capital of Germany •most populated city in Germany (3,438,000 people)
    Hamburg: •located in northern Germany •city in the Hamburg state •chief port on the Elbe River
    Munich: •located in southern Germany •city in the Bavaria state
    Cologne: •located in western Germany •city in the North Rhine-Westphalia state •chief port on the Rhine River

ENVIRONMENTAL/ECONOMIC:

Climate: temperate and marine; cool and wet; occasional foehn winds

Natural Resources: coal, lignite, natural gas, iron ore, copper
Agricultural Products: potatoes, wheat, barley, sugar beets, cattle
Major Exports: machinery, vehicles, chemicals, metals and manufactures, foodstuffs
Natural Hazards: floods

GHANA

Country Name: Republic of Ghana
Continent: Africa
Area: 92,100 sq mi / 238,537 sq km
Population: 24,966,000
Population Density: 271 people per sq mi / 105 people per sq km
Capital: Accra

PHYSICAL:

Highest Point: Mount Afadjato 2,904 ft / 885 m
Lowest Point: Gulf of Guinea 0 ft / 0 m (sea level)
Gulfs:
   Gulf of Guinea: •forms part of the southwestern coast of Ghana •feeds into the Atlantic Ocean
Bays:
   Bight of Benin: •forms most of the southern coast •feeds into the Gulf of Guinea
Lakes:
   Lake Volta: •covers parts of central and eastern Ghana •fed by the White Volta and Black Volta Rivers •dammed by the Akosombo Dam
Rivers:
   Black Volta: •tributary of the Volta River •forms part of Ghana's border with Cote d'Ivoire and Burkina Faso •extends into Burkina Faso
   Red Volta: •tributary of the White Volta River •extends into Burkina Faso
   Volta: •has its mouth in the Bight of Benin
   White Volta: •tributary of the Volta River •extends into Burkina Faso
Dams:
   Akosombo: •located in southeastern Ghana •on the Volta River •forms Lake Volta

POLITICAL:

Independence: March 6, 1957 (from the United Kingdom)
Former Name: Gold Coast
Bordering Countries: Togo, Cote d'Ivoire, Burkina Faso (3)
Administrative Divisions: Ashanti, Brong-Ahafo, Central, Eastern, Greater Accra, Northern, Upper East, Upper West, Volta, Western (10 regions)
Ethnic/Racial Groups: Akan, Moshi-Dagomba, Ewe, Ga, Gurma, Yoruba
Religions: Christian, indigenous beliefs, Islam
Languages: English, Akan, Moshi-Dagomba, Ewe, Ga
Currency: cedi

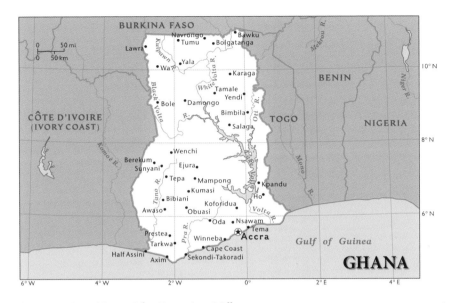

Current President: John Evans Atta Mills
Cities (capital, largest, or with at least a million people):
 Accra: •located in southern Ghana •city in the Greater Accra region •capital of Ghana •most populated city in Ghana (2,269,000 people) •chief port on the Bight of Benin
 Kumasi: •located in south-central Ghana •city in the Ashanti region •second largest city in Ghana •major business center

ENVIRONMENTAL/ECONOMIC:
Climate: tropical; warm and dry along the southeastern coast; hot and humid in the southwest; hot and dry in the north
Natural Resources: gold, timber, industrial diamonds, bauxite, manganese
Agricultural Products: cacao, rice, coffee, cassava (tapioca), timber
Major Exports: gold, cacao, timber, tuna, bauxite
Natural Hazards: harmattan winds, droughts

GREECE
Name: Hellenic Republic
Continent: Europe
Area: 50,949 sq mi / 131,957 sq km
Population: 11,329,000
Population Density: 222 people per sq mi / 86 people per sq km
Capital: Athens

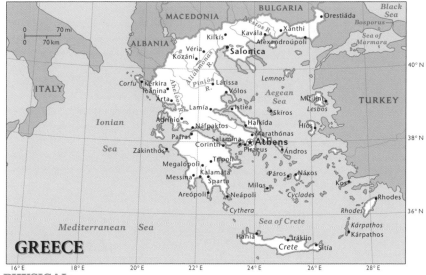

PHYSICAL:
Highest Point: Mount Olympus 9,570 ft / 2,917 m
Lowest Point: Mediterranean Sea 0 ft / 0 m (sea level)
Mountain Ranges:
    Pindus: •cover much of western Greece •extend into Albania
    Pirin: •located in northern Greece •extend into Bulgaria
    Rhodope: •located in northeastern Greece •extend into Bulgaria
Peaks (minimum elevation 8,000 ft / 2,400 m):
    Mount Olympus: •located in northern Greece •highest point in Greece
Seas:
    Aegean: •forms much of the eastern coast of Greece •feeds into the Sea of Crete
    Ionian: •forms much of the western coast of Greece •feeds into the Mediterranean Sea
    Mediterranean: •forms part of the western and southern coasts of Greece •feeds into the Atlantic Ocean
    Sea of Crete: •forms much of the southern coast of Greece •feeds into the Mediterranean Sea
Gulfs:
    Gulf of Corinth: •forms part of the western coast of Greece •feeds into the Gulf of Patrai
    Gulf of Euboea: •forms part of the central coast of Greece •feeds into the Aegean Sea
    Gulf of Patrai: •forms part of the western coast of Greece •feeds into the Ionian Sea
    Saronic: •forms part of the central coast of Greece •feeds into the Sea of Crete
    Thermaic: •forms part of the northern coast of Greece •feeds into the Aegean Sea

Lakes:

Lake Prespa: •located in northwestern Greece •on the border with Albania and Macedonia •on the edge of the Pindus Mountains
Rivers:
Evros: •has its mouth in the Aegean Sea •forms the Greece-Turkey border
Peninsulas:
Peloponnesus: •bordered by the Gulf of Corinth, Gulf of Patrai, Mediterranean Sea, Sea of Crete, and Saronic Gulf
Islands:
Cephalonia: •island off the western coast of Greece •bordered by the Ionian and Mediterranean Seas
Chios: •island off the coast of Turkey •surrounded by the Aegean Sea
Corfu: •island off the northwestern coast of Greece •surrounded by the Ionian Sea
Crete: •island between the Mediterranean Sea and the Sea of Crete
Kiklades (Cyclades): •island group bordered by the Aegean Sea and the Sea of Crete
Euboea: •island off the coast of central Greece •bordered by the Aegean Sea and the Gulf of Euboea
Lesvos: •island off the coast of Turkey •surrounded by the Aegean Sea
Limnos: •island off the coast of Turkey •surrounded by the Aegean Sea
Rhodes: •island off the coast of Turkey •bordered by the Sea of Crete and the Mediterranean Sea
Samos: •island off the coast of Turkey •bordered by the Aegean Sea
Thassos: •island off the northeastern coast of Greece •surrounded by the Aegean Sea

POLITICAL:
Independence: 1829 (from the Ottoman Empire)
Former Name: Kingdom of Greece
Bordering Countries: Bulgaria, Albania, Macedonia, Turkey (4)
Regions:
Macedonia: •region in northern Greece •extends into Macedonia and Bulgaria •east of the Pindus Mountains •south of the Pirin Mountains •north of the Thermaic Gulf and the Aegean Sea
Thessaly: •region in central Greece •east of the Pindus Mountains •west of the Thermaic Gulf and the Aegean Sea
Thrace: •region in northeastern Greece •extends into Bulgaria and Turkey •south of the Rhodope Mountains •north of the Aegean Sea
Administrative Divisions: Anatoliki Makedonia Kai Thraki, Atiki, Dytiki Elada, Dytiki Macedonia, Ionia Nisia, Ipiros, Kentriki Macedonia, Kriti, Notio Egeo, Peloponnisos, Sterea Elada, Thessalia, Voreio Egeo (13 administrative regions)
Ethnic/Racial Groups: predominantly Greek
Religion: Christianity (predominantly Greek Orthodox)
Language: predominantly Greek
Currency: euro
Current President: Karolos Papoulias
Cities (capital, largest, or with at least a million people):

Athens: •located in central Greece •city in the Attiki prefecture •capital of Greece •most populated city in Greece (3,252,000 people) •chief port on the Saronic Gulf

ENVIRONMENTAL/ECONOMIC:
Climate: temperate, mild and wet winters/hot and dry summers
Natural Resources: lignite, petroleum, iron ore, bauxite, lead
Agricultural Products: wheat, corn, barley, sugar beets, beef
Major Exports: food and beverages, manufactured goods, petroleum products, chemicals
Natural Hazards: earthquakes

GRENADA
Country Name: Grenada
Continent: North America
Area: 133 sq mi / 344 sq km
Population: 105,000
Population Density: 789 people per sq mi / 305 people per sq km
Capital: Saint George's

PHYSICAL:
Highest Point: Mount Saint Catherine 2,756 ft / 840 m
Lowest Point: Caribbean Sea 0 ft / 0 m (sea level)
Oceans:
   Atlantic: •forms the eastern coast of Grenada
Seas:
   Caribbean: •forms the western coast of Grenada •feeds into the Atlantic Ocean
Islands:
   Grenada: •largest island in Grenada •bordered by the Atlantic Ocean and the Caribbean Sea

POLITICAL:
Independence: February 7, 1974 (from the United Kingdom)
Administrative Divisions: Saint Andrew, Saint David, Saint George, Saint John, Saint Mark, Saint Patrick (6 parishes) •Carriacou and Petit Martinique (1 dependency)
Ethnic/Racial Groups: black, Creole
Religion: Christianity (mainly Roman Catholic, Anglican, other Protestant)
Languages: English, French patois
Currency: East Caribbean dollar
Major Exports: bananas, cacao, nutmeg, fruits and vegetables
Current Prime Minister: Tillman Thomas

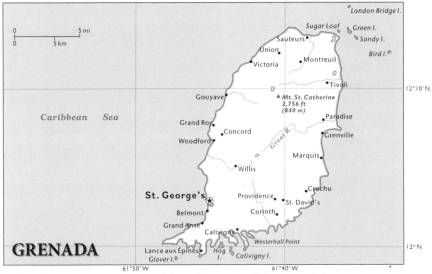

GRENADA

Cities (capital, largest, or with at least a million people):
Saint George's: •located on the island of Grenada •city in the Saint George parish •capital of Grenada •most populated city in Grenada (40,000 people) •chief port on the Caribbean Sea

ENVIRONMENTAL/ECONOMIC:
Climate: tropical, moderated by trade winds
Natural Resources: timber, tropical fruit, deepwater harbors
Agricultural Products: bananas, cacao, nutmeg, mace
Natural Hazards: hurricanes

GUATEMALA
Country Name: Republic of Guatemala
Continent: North America
Area: 42,042 sq mi / 108,889 sq km
Population: 14,740,000
Population Density: 351 people per sq mi / 135 people per sq km
Capital: Guatemala City

PHYSICAL:
Highest Point: Volcan Tajumulco 13,816 ft / 4,211 m
Lowest Point: Pacific Ocean 0 ft / 0 m (sea level)
Mountain Ranges:
Maya: •located in eastern Guatemala •extend into Belize

Sierra Madre: •cover much of southern Guatemala •extend into Mexico
Peaks (minimum elevation 8,000 ft / 2,400 m):
Volcan Tajumulco: •located in the Sierra Madre •highest point in Guatemala
Oceans:
Pacific: •forms the southern coast of Guatemala
Gulfs:
Gulf of Honduras: •forms the eastern coast of Guatemala •feeds into the Caribbean Sea
Lakes:
Lake Izabal: •located in eastern Guatemala

POLITICAL:
Independence: September 15, 1821 (from Spain)
Bordering Countries: Mexico, Belize, Honduras, El Salvador (4)
Administrative Divisions: Alta Verapaz, Baja Verapaz, Chimaltenango, Chiquimula, El Progreso, Escuintla, Guatemala, Huehuetenango, Izabal, Jalapa, Jutiapa, Peten, Quetzaltenango, Quiche, Retalhuleu, Sacatepequez, San Marcos, Santa Rosa, Solola, Suchitepequez, Totonicapan, Zacapa (22 departments)
Ethnic/Racial Groups: mestizo, Amerindian
Religions: Christianity (Roman Catholic, Protestant), indigenous Mayan beliefs
Languages: Spanish, Amerindian languages
Currency: quetzal, U.S. dollar, others allowed
Current President: Alvaro Colom Caballeros
Cities (capital, largest, or with at least a million people):
Guatemala City: •located in southern Guatemala •city in the Guatemala department

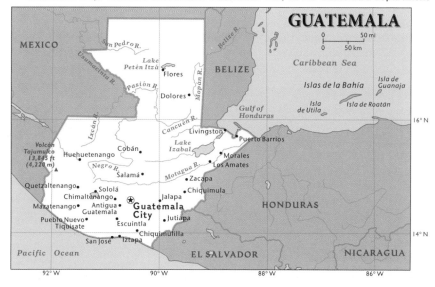

• capital of Guatemala • most populated city in Guatemala (1,075,000 people) • located in a valley in the Sierra Madre

ENVIRONMENTAL/ECONOMIC:
Climate: tropical; hot and humid in the lowlands, cooler in the highlands
Natural Resources: petroleum, nickel, rare woods, fish, chicle
Agricultural Products: sugarcane, corn, bananas, coffee, cattle
Major Exports: coffee, sugar, bananas, fruits and vegetables, cardamom
Natural Hazards: volcanoes, earthquakes, hurricanes

GUINEA
Country Name: Republic of Guinea
Continent: Africa
Area: 94,926 sq mi / 245,857 sq km
Population: 10,232,000
Population Density: 108 people per sq mi / 42 people per sq km
Capital: Conakry

PHYSICAL:
Highest Point: Mont Nimba 5,748 ft / 1,752 m
Lowest Point: Atlantic Ocean 0 ft / 0 m (sea level)
Grasslands/Prairies:
  Sahel: • covers much of northern Guinea • extends into Guinea-Bissau, Senegal, Mali, and Cote d'Ivoire
Oceans:
  Atlantic: • forms the western coast of Guinea
Rivers:
  Bafing: • has its source in central Guinea • forms part of the Guinea-Mali border • extends into Mali
  Gambie (Gambia): • has its source in northern Guinea • extends into Senegal
  Niger: • has its source in southern Guinea • extends into Mali
Regions:
  Fouta Djallon: • highland region that covers parts of northern and central Guinea • south of the Sahel

POLITICAL:
Independence: October 2, 1958 (from France)
Former Name: French Guinea
Bordering Countries: Mali, Sierra Leone, Cote d'Ivoire, Liberia, Guinea-Bissau, Senegal (6)

Administrative Divisions: Beyla, Boffa, Boke, Coyah, Dabola, Dalaba, Dinguiraye, Dubreka, Faranah, Forecariah, Fria, Gaoual, Gueckedou, Kankan,

Kerouane, Kindia, Kissidougou, Koubia, Koundara, Kouroussa, Labe, Lelouma, Lola, Macenta, Mali, Mamou, Mandiana, Nzerekore, Pita, Siguiri, Telimele, Tougue, Yomou (33 prefectures) •Conakry (1 special zone)
Ethnic/Racial Groups: Peuhl, Malinke, Soussou
Religions: Islam, Christianity, indigenous beliefs
Languages: French, local languages
Currency: Guinean franc
Current President: Alpha Conde
Cities (capital, largest, or with at least a million people):
  Conakry: •located in western Guinea •city in the Conakry special zone •capital of Guinea •most populated city in Guinea (1,597,000 people) •chief port on the Atlantic Ocean

ENVIRONMENTAL/ECONOMIC:
Climate: hot and humid with rainy/dry seasons; harmattan winds
Natural Resources: bauxite, iron ore, diamonds, gold, uranium
Agricultural Products: rice, coffee, pineapples, palm kernels
Major Exports: bauxite, alumina, gold, diamonds, coffee
Natural Disasters: harmattan haze

GUINEA-BISSAU

Country Name: Republic of Guinea-Bissau
Continent: Africa
Area: 13,948 sq mi / 36,125 sq km
Population: 1,610,000
Population Density: 115 people per sq mi / 45 people per sq km
Capital: Bissau

PHYSICAL:
Highest Point: unnamed location 984 ft / 300 m
Lowest Point: Atlantic Ocean 0 ft / 0 m (sea level)
Grasslands/Prairies:
   Sahel: •covers most of the country •extends into Senegal and Guinea
Oceans:
   Atlantic: •forms the western coast of Guinea-Bissau

POLITICAL:
Independence: September 24, 1973 (from Portugal)
Former Name: Portuguese Guinea
Bordering Countries: Senegal, Guinea (2)
Administrative Divisions: Bafata, Biombo, Bissau, Bolama, Cacheu, Gabu, Oio, Quinara, Tombali (9 regions)
Ethnic/Racial Groups: Balanta, Fula, Manjaca, Mandinga, Papel
Religions: indigenous beliefs, Islam, Christianity
Languages: Portuguese, Crioulo, African languages
Currency: Communaute Financiere Africaine franc

Current President: Malam Bacai Sanha
Cities (capital, largest, or with at least a million people):
  Bissau: •located in western Guinea-Bissau •city in the Bissau region •capital of Guinea-Bissau •most populated city in Guinea-Bissau (302,000 people) •chief port on the Atlantic Ocean

ENVIRONMENTAL/ECONOMIC:
Climate: tropical, hot and humid, rainy/dry seasons, harmattan winds
Natural Resources: fish, timber, phosphates, bauxite, petroleum
Agricultural Products: rice, corn, beans, cassava (tapioca), timber
Major Exports: cashew nuts, shrimp, peanuts, palm kernels, sawn lumber
Natural Hazards: harmattan haze, brush fires

GUYANA
Country Name: Co-operative Republic of Guyana
Continent: South America
Area: 83,000 sq mi / 214,969 sq km
Population: 757,000
Population Density: 9 people per sq mi / 3.5 people per sq km
Capital: Georgetown

PHYSICAL:
Highest Point: Mount Roraima 8,986 ft / 2,739 m
Lowest Point: Atlantic Ocean 0 ft / 0 m (sea level)
Mountain Ranges:
  Guiana Highlands: •cover much of central and southern Guyana •extend into Venezuela, Brazil, and Suriname
Peaks (minimum elevation 8,000 ft / 2,400 m):
  Mount Roraima: •located in the Guiana Highlands •on the border with Venezuela and Brazil •highest point in Guyana
Oceans:
  Atlantic: •forms the northern coast of Guyana
Rivers:
  Essequibo: •has its mouth in the Atlantic Ocean •has its source in the Guiana Highlands

POLITICAL:
Independence: May 26, 1966 (from the United Kingdom)
Former Name: British Guiana
Bordering Countries: Brazil, Venezuela, Suriname (3)

Administrative Divisions: Barima-Waini, Cuyuni-Mazaruni, Demerara-Mahaica, East Berbice–Corentyne, Essequibo Islands-West Demerara, Mahaica-Berbice, Pomeroon-Supenaam, Potaro-Siparuni, Upper Demerara–Berbice, Upper Takutu–Upper Essequibo (10 regions)
Ethnic/Racial Groups: East Indian, black, Amerindian
Religions: Christianity, Hinduism, Islam
Languages: English, Amerindian dialects, Creole, Hindi, Urdu
Currency: Guyanese dollar
Current President: Bharrat Jagdeo
Cities (capital, largest, or with at least a million people):
Georgetown: •located in northeastern Guyana •city in the Demerara-Mahaica region •capital of Guyana •most populated city in Guyana (132,000 people) •chief port on the Atlantic Ocean

ENVIRONMENTAL/ECONOMIC:
Climate: tropical; hot and humid with two rainy seasons; moderated by trade winds
Natural Resources: bauxite, gold, diamonds, hardwood timber, shrimp
Agricultural Products: sugar, rice, wheat, vegetable oils, beef
Major Exports: sugar, gold, bauxite/alumina, rice, shrimp
Natural Hazards: flash floods

HAITI

Country Name: Republic of Haiti
Continent: North America
Area: 10,714 sq mi / 27,750 sq km
Population: 10,124,000
Population Density: 945 people per sq mi / 365 people per sq km
Capital: Port-au-Prince

PHYSICAL:

Highest Point: Chaine de la Selle 8,793 ft / 2,680 m
Lowest Point: Caribbean Sea 0 ft / 0 m (sea level)
Peaks (minimum elevation of 8,000 ft / 2,400 m):
  Chaine de la Selle: •located in southeastern Haiti •highest point in Haiti
Oceans:
  Atlantic: •forms much of the northern coast of Haiti
Seas:
  Caribbean: •forms the southern coast of Haiti •feeds into the Atlantic Ocean

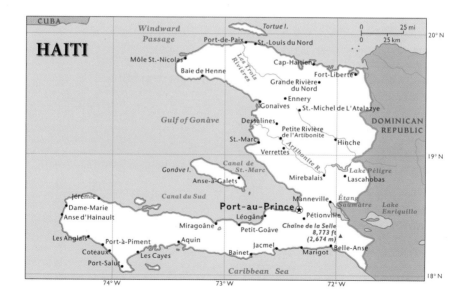

Gulfs:
　Golfe de la Gonave: •forms the west coast of Haiti
Passages:
　Windward: •forms part of the northwestern coast of Haiti •separates Haiti from Cuba •connects the Atlantic Ocean and the Caribbean Sea
Islands:
　Hispaniola: •Haiti takes up the western portion of this island •bordered by the Atlantic Ocean, Caribbean Sea, Golfe de la Gonave, and Windward Passage •the Dominican Republic takes up the eastern portion of the island
　Ile de la Gonave: •island off the west coast of Hispaniola •surrounded by the Golfe de la Gonave

POLITICAL:
Independence: January 1, 1804 (from France)
Bordering Country: Dominican Republic (1)
Administrative Divisions: Artibonite, Centre, Grand Anse, Nord, Nord-Est, Nord-Ouest, Ouest, Sud, Sud-Est (9 departments)
Ethnic/Racial Groups: black, Creole and other
Religions: Christianity (Roman Catholic, Protestant), voodoo traditions
Languages: French, Creole
Currency: gourde
Current Prime Minister: Gary Conille
Cities (capital, largest, or with at least a million people):
　Port-au-Prince: •located on the island of Hispaniola •city in the Ouest department •capital of Haiti •most populated city in Haiti (2,643,000 people) •chief port on the Golfe de la Gonave

ENVIRONMENTAL/ECONOMIC:
Climate: tropical; semi-arid in the eastern mountains
Natural Resources: bauxite, copper, calcium carbonate, gold, marble
Agricultural Products: coffee, mangoes, sugarcane, rice, wood
Major Exports: manufactures, coffee, oils, cocoa
Natural Hazards: hurricanes, floods, earthquakes, droughts

HONDURAS
Country Name: Republic of Honduras
Continent: North America
Area: 43,433 sq mi / 112,492 sq km
Population: 7,755,000
Population Density: 179 people per sq mi / 69 people per sq km
Capital: Tegucigalpa

PHYSICAL:
Highest Point: Cerro Las Minas 9,347 ft / 2,849 m
Lowest Point: Caribbean Sea 0 ft / 0 m (sea level)
Peaks (minimum elevation of 8,000 ft / 2,400 m):
    Las Minas: •located in western Honduras •highest point in Honduras
Seas:
    Caribbean: •forms most of the northern coast of Honduras •feeds into the Atlantic Ocean
Gulfs:
    Golfo de Fonseca: •forms the southern coast of Honduras •feeds into the Pacific Ocean
    Gulf of Honduras: •forms part of the northwestern coast of Honduras •feeds into the Caribbean Sea
Islands:
    Islas de la Bahia: •island group off the northern coast of Honduras •surrounded by the Caribbean Sea

POLITICAL:
Independence: September 15, 1821 (from Spain)
Bordering Countries: Nicaragua, El Salvador, Guatemala (3)
Administrative Divisions: Atlantida, Choluteca, Colon, Comayagua, Copan, Cortes, El Paraiso, Francisco Morazan, Gracias a Dios, Intibuca, Islas de la Bahia, La Paz, Lempira, Ocotepeque, Olancho, Santa Barbara, Valle, Yoro (18 departments)
Ethnic/Racial Groups: mestizo, Amerindian, black, white

Religion: Christianity (predominantly Roman Catholic)
Languages: Spanish, Amerindian dialects
Currency: lempira
Current President: Porfirio Lobo Sosa
Cities (capital, largest, or with at least a million people):
  Tegucigalpa: •located in central Honduras •city in the Francisco Morazan department •capital of Honduras •most populated city in Honduras (1,000,000 people)

ENVIRONMENTAL/ECONOMIC:
Climate: subtropical in the lowlands; temperate in the mountains
Natural Resources: timber, gold, silver, copper, lead
Agricultural Products: bananas, coffee, citrus, beef, timber
Major Exports: coffee, bananas, shrimp, lobster, meat
Natural Hazards: earthquakes, hurricanes, floods

HUNGARY
Country Name: Republic of Hungary
Continent: Europe
Area: 35,919 sq mi / 93,030 sq km
Population: 9,972,000
Population Density: 278 people per sq mi / 107 people per sq km
Capital: Budapest

PHYSICAL:
Highest Point: Kekes 3,330 ft / 1,015 m
Lowest Point: Tisza River 256 ft / 78 m
Grasslands/Prairies:
> Great Alfold: •covers an area from southern to northeastern Hungary •extends into Croatia, Serbia and Montenegro, Romania, Slovakia, and Ukraine

Lakes:
> Balaton: •located in western Hungary

Rivers:
> Danube: •forms part of the Hungary-Slovakia border •extends into Austria, Serbia and Montenegro, and Romania
> Drava: •forms part of the Hungary-Croatia border •extends into Croatia and Slovenia
> Tisza: •extends into Ukraine and into Serbia and Montenegro

POLITICAL:
Independence: 1001
Bordering Countries: Slovakia, Romania, Austria, Croatia, Serbia, Ukraine, Slovenia (7)
Administrative Divisions: Bacs-Kiskun, Baranya, Bekes, Borsod-Abauj-Zemplen, Csongrad, Fejer, Gyor-Moson-Sopron, Hajdu-Bihar, Heves, Jasz-Nagykun-Szolnok, Komarom-Esztergom, Nograd, Pest, Somogy, Szabolcs-Szatmar-Bereg, Tolna, Vas, Veszprem, Zala (19 counties) •Bekescsaba, Debrecen, Dunaujvaros, Eger, Gyor, Hodmezovasarhely, Kaposvar, Kecskemet, Miskolc, Nagykanizsa, Nyiregyhaza, Pecs, Sopron, Szeged, Szekesfehervar, Szolnok, Szombathely, Tatabanya, Veszprem, Zalaegerszeg (20 urban counties) •Budapest (1 capital city division)
Ethnic/Racial Groups: Hungarian, Roma, German, Serb
Religion: Christianity (Roman Catholic, Calvinist, Lutheran)
Language: Hungarian
Currency: forint
Current President: Pal Schmitt
Cities (capital, largest, or with at least a million people):
> Budapest: •located in northern Hungary •city in the Budapest capital city division •capital of Hungary •most populated city in Hungary (1,705,000 people) •chief port on the Danube River

ENVIRONMENTAL/ECONOMIC:
Climate: temperate, cold and humid winters/warm summers
Natural Resources: bauxite, coal, natural gas, fertile soils, arable land
Agricultural Products: wheat, corn, sunflower seed, potatoes, sugar beets
Major Exports: machinery and equipment, other manufactures, food products, raw materials
Natural Hazards: floods, thunderstorms

ICELAND

Country Name: Republic of Iceland
Continent: Europe
Area: 39,769 sq mi / 103,000 sq km
Population: 319,000
Population Density: 8 people per sq mi / 3 people per sq km (least dense in Europe)
Capital: Reykjavik

PHYSICAL:

Highest Point: Hvannadalshnukur 6,952 ft / 2,119 m
Lowest Point: Atlantic Ocean 0 ft / 0 m (sea level)
Oceans:
  Atlantic: •forms most of the southern and eastern coasts of Iceland •forms part of the western coast of Iceland
Seas:
  Greenland: •forms most of the northern coast of Iceland •feeds into the Atlantic and Arctic Oceans

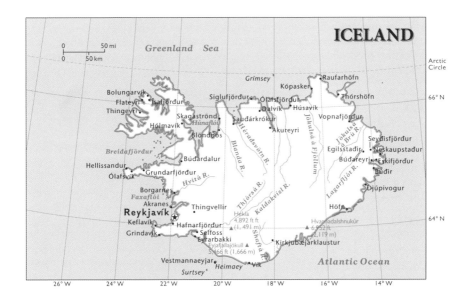

Straits:
> Denmark: •forms part of the northwestern coast of Iceland •separates Iceland from Greenland (territory of Denmark) •connects the Atlantic Ocean and the Greenland Sea

Islands:
> Iceland: •main island in Iceland •bordered by the Atlantic Ocean, Greenland Sea, and Denmark Strait
> Surtsey: •island off southern coast of Iceland •formed by volcanic activity from 1963 to 1967

POLITICAL:
Independence: June 17, 1944 (from Denmark)
Administrative Divisions: Austurland, Hofudhborgarsvaedhi, Nordhurland Eystra, Nordhurland Vestra, Sudhurland, Suournes, Vestfiroir, Vesturland (8 regions)
Ethnic/Racial Groups: Icelandic
Religion: Christianity (primarily Evangelical Lutheran)
Languages: Icelandic, English, Nordic languages, German
Currency: Icelandic krona
Current President: Olafur Ragnar Grimsson
Cities (capital, largest, or with at least a million people):
> Reykjavik: •located in southwestern Iceland •city in the Hofudhborgarsvaedhi region •capital of Iceland •most populated city in Iceland (198,000 people) •chief port on the Atlantic Ocean

ENVIRONMENTAL/ECONOMIC:
Climate: temperate; moderated by the North Atlantic Current; mild winters/cool summers
Natural Resources: fish, hydropower, geothermal power, diatomite
Agricultural Products: potatoes, green vegetables, chicken, pork, fish
Major Exports: fish and fish products, animal products, aluminum, diatomite, ferrosilicon
Natural Hazards: earthquakes, volcanoes

INDIA
Country Name: Republic of India
Continent: Asia
Area: 1,269,221 sq mi / 3,287,270 sq km
Population: 1,241,275,000
Population Density: 978 people per sq mi / 378 people per sq km
Capital: New Delhi

PHYSICAL:
Highest Point: Kanchenjunga 28,209 ft / 8,598 m
Lowest Point: Indian Ocean 0 ft / 0 m (sea level)
Mountain Ranges:
> Eastern Ghats: •border the eastern edge of the Deccan Plateau

Himalaya: •located in northern India •extend into Pakistan, China, Nepal, and Bhutan
Karakoram Range: •located in northwestern India •extends into Pakistan and China
Satpura Range: •located in central India •north of the Deccan Plateau
Western Ghats: •border the western edge of the Deccan Plateau

Peaks:
Kamet: •located in the Himalaya •on the India-China border
Kanchenjunga: •located in the Himalaya •on the India-Nepal border •highest point in India
Nanda Devi: •located in the Himalaya

Passes:
Karakoram: •located in the Karakoram Range •on the India-China border

Grasslands/Prairies:
Sundarbans: •located in eastern India •north of the Bay of Bengal •extend into Bangladesh

Plateau Regions:
Chota Nagpur Plateau: •located in central India •plateau at the eastern end of the Satpura Range
Deccan Plateau: •covers much of southern India and part of central India •basin between the Satpura Range, Eastern Ghats, and Western Ghats

Deserts:
Great Indian: •located in northwestern India •extends into Pakistan

Oceans:
Indian: •forms most of the southern coast of India

Seas:
    Andaman: •forms part of the southeastern coast of India •feeds into the Indian Ocean
    Arabian: •forms part of the western coast of India •feeds into the Indian Ocean
    Laccadive: •forms part of the southwestern coast of India •feeds into the Indian Ocean

Gulfs:
    Gulf of Khambhat: •forms part of the western coast of India •feeds into the Arabian Sea
    Gulf of Kutch: •forms part of the western coast of India •feeds into the Arabian Sea
    Gulf of Mannar: •forms part of the southern coast of India •feeds into the Indian Ocean

Bays:
    Bay of Bengal: •forms most of the eastern coast of India •feeds into the Indian Ocean

Straits:
    Palk: •forms part of the southern coast of India •separates India from Sri Lanka •connects the Bay of Bengal and the Gulf of Mannar

Rivers:
    Brahmaputra: •located in northeastern India •extends into China and Bangladesh
    Ganges: •has its source in the Himalaya •extends into Bangladesh
    Godavari: •has its mouth in the Bay of Bengal •has its source in the Western Ghats
    Indus: •extends into Pakistan and China
    Narmada: •has its mouth in the Gulf of Khambhat •has its source in the Satpura Range
    Yamuna: •tributary of the Ganges River •has its source in the Himalaya

Deltas:
    Mouths of the Ganges: •mouth of the Ganges River •feeds into the Bay of Bengal •extends into Bangladesh

Peninsulas:
    Kathiawar: •bordered by the Arabian Sea, Gulf of Kutch, and Gulf of Khambhat

Islands:
    Andaman: •island group between the Andaman Sea and the Bay of Bengal
    Lakshadweep: •island group between the Laccadive Sea and the Indian Ocean
    Nicobar: •island group between the Andaman Sea and the Indian Ocean

POLITICAL:
Independence: August 15, 1947 (from the United Kingdom)
Bordering Countries: Bangladesh, China, Pakistan, Nepal, Myanmar, Bhutan (6)
Administrative Divisions: Andhra Pradesh, Arunachal Pradesh, Assam, Bihar, Chhattisgarh, Goa, Gujarat, Haryana, Himachal Pradesh, Jammu and Kashmir, Jharkhand, Karnataka, Kerala, Madhya Pradesh, Maharashtra, Manipur, Meghalaya, Mizoram, Nagaland, Orissa, Punjab, Rajasthan, Sikkim, Tamil Nadu, Tripura, Uttarakhand, Uttar Pradesh, West Bengal (28 states) •Andaman and Nicobar Islands, Chandigarh, Dadra and Nagar Haveli, Daman and Diu, Delhi, Lakshadweep, Puducherry (7 union territories)
Ethnic/Racial Groups: Indo-Aryan, Dravidian
Religions: Hinduism, Islam, Christianity, Sikh, Buddhism, Jain, Parsi
Languages: Hindi, English, 14 other official languages

Currency: Indian rupee
Current President: Pratibha Patil
Cities (capital, largest, or with at least 1.5 million people):
  Delhi: •located in northwestern India •city in the Delhi union territory •most populated city in India (22,157,000 people) •chief port on the Yamuna River
  Mumbai (Bombay): •located in western India •city in the Maharashtra state •chief port on the Arabian Sea
  Kolkata (Calcutta): •located in eastern India •city in the West Bengal state •north of the Sundarbans •on the edge of the Mouths of the Ganges
  Bangalore: •located in southwestern India •city in the Karnataka state •on the edge of the Eastern Ghats on the Deccan Plateau
  Chennai (Madras): •located in southern India •city in the Tamil Nadu state •chief port on the Bay of Bengal
  Ahmadabad: •located in western India •city in the Gujarat state
  Hyderabad: •located in central India •city in the Andhra Pradesh state •on the Deccan Plateau
  Pune: •located in western India •city in the Maharashtra state •on the edge of the Western Ghats
  Kanpur: •located in central India •city in the Uttar Pradesh state •chief port on the Ganges River
  Nagpur: •located in central India •city in the Maharashtra state •on the edge of the Satpura Range on the Deccan Plateau
  Surat: •located in western India •city in the Gujarat state •chief port on the Gulf of Khambhat
  Jaipur: •located in northwestern India •city in the Rajasthan state
  Lucknow: •located in northern India •city in the Uttar Pradesh state
  Indore: •located in central India •city in the Madhya Pradesh state
  Bhophal: •located in central India •city in the Madhya Pradesh state
  Ludhiana: •located in northern India •city in the Punjab state
  Patna: • located in northern India •city in the Bihar state
  Vadodara: •located in western India •city in the Gujarat state
  Agra: •located in northern India •city in the Uttar Pradesh state
  New Delhi: •located in northwestern India •city in the Delhi union territory •on Yamuna River •capital of India (295,000 people)

ENVIRONMENTAL/ECONOMIC:
Climate: tropical monsoon in south; temperate in north
Natural Resources: coal, iron ore, manganese, mica, bauxite
Agricultural Products: rice, wheat, oilseed, cotton, cattle
Major Exports: textile goods, gems and jewelry, engineering goods, chemicals, leather manufactures
Natural Hazards: droughts, flash floods, monsoonal floods, severe thunderstorms, earthquakes

INDONESIA

Country Name: Republic of Indonesia
Continent: Asia
Area: 742,308 sq mi / 1,922,570 sq km
Population: 238,181,000
Population Density: 321 people per sq mi / 124 people per sq km
Capital: Jakarta

PHYSICAL:
Highest Point: Puncak Jaya 16,503 ft / 5,030 m
Lowest Point: Indian Ocean 0 ft / 0 m (sea level)
Mountain Ranges:
  Maoke: •located on the island of New Guinea •extend into Papua New Guinea
Peaks:
  Puncak Jaya: •located in the Maoke Mountains •highest point in Indonesia
Oceans:
  Indian: •forms part of the western and southern coasts of Indonesia
  Pacific: •forms part of the northern coast of New Guinea
Seas:
  Andaman: •forms part of the northern coast of Sumatra •feeds into the Indian Ocean
  Arafura: •forms part of the southeastern coast of Indonesia •feeds into the Timor and Coral Seas
  Bali: •forms part of the inner coast of Indonesia •feeds into the Java Sea
  Banda: •forms part of the inner coast of Indonesia •feeds into the Timor Sea

Celebes: •forms part of the northern coast of Indonesia •feeds into the Philippine Sea
Ceram: •forms part of the inner coast of Indonesia •feeds into the Halmahera Sea
Flores: •forms part of the inner coast of Indonesia •feeds into the Banda Sea
Halmahera: •forms part of the inner coast of Indonesia •feeds into the Pacific Ocean
Java: •forms part of the inner coast of Indonesia •feeds into the Indian Ocean
Molucca: •forms part of the inner coast of Indonesia •feeds into the Pacific Ocean
Philippine: •forms part of the northeastern coast of Indonesia •feeds into the Pacific Ocean
Savu: •forms part of the southern coast of Indonesia •feeds into the Indian Ocean
South China: •forms part of the western coast of Indonesia •feeds into the Java and Philippine Seas
Timor: •forms part of the southern coast of Indonesia •feeds into the Indian Ocean

Gulfs:
Gulf of Bone: •forms part of the inner coast of Celebes •feeds into the Flores Sea
Gulf of Tomini: •forms part of the inner coast of Celebes •feeds into the Molucca Sea

Straits:
Makassar: •forms part of the inner coast of Indonesia •separates the island of Borneo from the island of Celebes •connects the Celebes and Java Seas
Strait of Malacca: •forms part of the northern coast of Sumatra •separates the island of Sumatra from Thailand and Malaysia •connects the South China and Andaman Seas
Sumba: •forms part of the southern coast of Indonesia •separates the islands of Sumbawa and Flores from the island of Sumba •connects the Indian Ocean and the Savu Sea
Sunda: •forms part of the southwestern coast of Indonesia •separates the island of Sumatra from the island of Java •connects the Indian Ocean and the Java Sea

Peninsulas:
Minahasa: •bordered by the Celebes Sea, Molucca Sea, and Gulf of Tomimi •located on the island of Celebes
Doberai: •bordered by the Pacific Ocean and Ceram Sea •located on the northwestern part of the island of New Guinea

Islands:
Bali: •westernmost of the Lesser Sunda Islands •bordered by the Bali Sea and Indian Ocean
Borneo: •second largest island in Indonesia •bordered by the South China Sea, Java Sea, Celebes Sea, and Makassar Strait •parts of the island belong to Malaysia and Brunei
Buru: •one of the Moluccas •bordered by the Ceram and Banda Seas
Sulawesi (Celebes): •easternmost of the Greater Sunda Islands •bordered by the Celebes Sea, Flores Sea, Molucca Sea, Banda Sea, Gulf of Tomini, Gulf of Bone, and Makassar Strait
Ceram: •one of the Moluccas •bordered by the Ceram and Banda Seas
Flores: •one of the Lesser Sunda Islands •bordered by the Flores and Savu Seas
Halmahera: •northernmost of the Moluccas •bordered by the Pacific Ocean, the Molucca Sea, and the Halmahera Sea

Java: •southernmost of the Greater Sunda Islands •bordered by the Indian Ocean, Java Sea, and Sunda Strait
Lombok: •one of the Lesser Sunda Islands •bordered by the Bali Sea and Indian Ocean
New Guinea: •largest island in Indonesia •bordered by the Pacific Ocean, Ceram Sea, and Arafura Sea •the country of Papua New Guinea makes up the eastern half
Sumatra: •westernmost of the Greater Sunda Islands •bordered by the Indian Ocean, Andaman Sea, South China Sea, Java Sea, Strait of Malacca, and Sunda Strait
Sumba: •one of the Lesser Sunda Islands •bordered by the Indian Ocean, Savu Sea, and Sumba Strait
Sumbawa: •one of the Lesser Sunda Islands •bordered by the Flores Sea and Indian Ocean
Timor: •easternmost of the Lesser Sunda Islands •bordered by the Savu and Timor Seas •the country of East Timor makes up the northern half of the island

POLITICAL:
Independence: August 17, 1945 (from the Netherlands)
Former Names: Netherlands East Indies, Dutch East Indies
Bordering Countries: Malaysia, Papua New Guinea, East Timor (Timor-Leste) (3)
Regions:
Irian Jaya: •region covering the western end of New Guinea •west of Papua New Guinea •south of the Pacific Ocean •east of the Ceram Sea •north of Arafura Sea
Administrative Divisions: Bali, Banten, Bengkulu, Gorontalo, Jambi, Jawa Barat, Jawa Tengah, Jawa Timur, Kalimantan Barat, Kalimantan Selatan, Kalimantan Tengah, Kalimantan Timur, Kepulauan Bangka Belitung, Kepulauan Riau, Lampung, Maluku, Maluku Utara, Nusa Tenggara Barat, Nusa Tenggara Timur, Papua Barat, Papua, Riau, Sulawesi Barat, Sulawesi Selatan, Sulawesi Tengah, Sulawesi Tenggara, Sulawesi Utara, Sumatera Barat, Sumatera Selatan, Sumatera Utara (30 provinces) •Aceh, Yogyakarta (2 special regions) •Jakarta Raya (1 special capital city district)
Ethnic/Racial Groups: Javanese, Sundanese, Madurese, Malay
Religions: Islam, Christianity (Protestant, Roman Catholic), Hinduism, Buddhism
Languages: Bahasa Indonesia, English, Dutch, Javanese, other local dialects
Currency: Indonesian rupiah
Current President: Susilo Bambang Yudhoyono
Cities (capital, largest, or with at least a million people):
Jakarta: •located on the island of Java •city in the Jakarta Raya special capital city district •capital of Indonesia •most populated city in Indonesia (9,121,000 people) •chief port on the Java Sea
Surabaya: •located on the island of Java •city in the Jawa Timur province •chief port on the Bali Sea
Bandung: •located on the island of Java •city in the Jawa Barat province
Medan: •located on the island of Sumatra •city in the Sumatera Utara province
Semarang: •located on the island of Java •city in the Jawa Tengah province

Palembang: •located on the island of Sumatra •city in the Sumatera Selatan province
Tangerang: •located on the island of Java •city in the Banten province
Makasar: •located on the island of Sulawesi •city in the Sulawesi Selatan province

ENVIRONMENTAL/ECONOMIC:
Climate: tropical; hot and humid; moderate in the highlands
Natural Resources: petroleum, tin, natural gas, nickel, timber
Agricultural Products: rice, cassava (tapioca), peanuts, rubber, poultry
Major Exports: oil and gas, electrical appliances, plywood, textiles, rubber
Natural Hazards: floods, droughts, tsunamis, earthquakes, volcanoes, forest fires

IRAN

Country Name: Islamic Republic of Iran
Continent: Asia
Area: 636,296 sq mi / 1,648,000 sq km
Population: 77,891,000
Population Density: 122 people per sq mi / 47 people per sq km
Capital: Tehran

PHYSICAL:
Highest Point: Qolleh-ye Damavand 18,606 ft / 5,671 m
Lowest Point: Caspian Sea 92 ft / 28 m below sea level

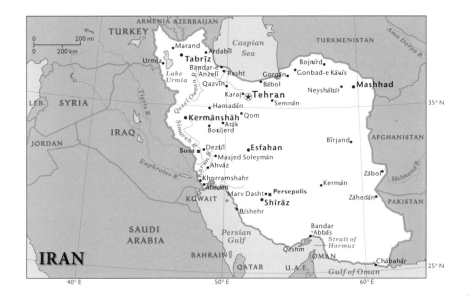

Mountain Ranges:
  Elburz: •located in northern Iran
  Zagros: •cover an area from northwestern to southeastern Iran •extend into Turkey and Iraq
Peaks:
  Qolleh-ye Damavand: •located in the Elburz Mountains •highest point in Iran
Depressions:
  Caspian Sea: •located in northern Iran •lowest point in Iran
Deserts:
  Dasht-e Kavir: •located in northern Iran
  Dasht-e Lut: •covers much of eastern Iran
Gulfs:
  Gulf of Oman: •forms the southeastern coast of Iran •feeds into the Arabian Sea
  Persian: •forms the southwestern coast of Iran •feeds into the Gulf of Oman
Straits:
  Strait of Hormuz: •forms part of the southern coast of Iran •separates Iran from Oman •connects the Persian Gulf and Gulf of Oman
Lakes:
  Caspian Sea: •forms the northern coast of Iran •also borders Azerbaijan and Turkmenistan •largest lake in the world
  Lake Urmia: •located in northwestern Iran •surrounded by the Zagros Mountains
Rivers:
  Aras: •forms part of Iran's border with Azerbaijan and Armenia •extends into Turkey and Azerbaijan

POLITICAL:
Independence: April 1, 1979
Former Name: Persia
Bordering Countries: Iraq, Afghanistan, Turkmenistan, Pakistan, Azerbaijan, Turkey, Armenia (7)
Regions:
  Baluchistan: •region in southeastern Iran •located north of the Gulf of Oman •extends into Pakistan
  Fars: •region in southern Iran •located in the southern Zagros Mountains •north of the Persian Gulf
  Khorasan: •region in northeastern Iran •located south of the Elburz Mountains in the Dasht-e Kavir
  Kurdistan: •region in northwestern Iran •located in the Zagros Mountains •extends into Turkey and Iraq
Administrative Divisions: Alborz, Ardabil, Bushehr, Chahar Mahall and Bakhtiari, East Azerbaijan, Esfahan, Fars, Gilan, Golestan, Hamadan, Hormozgan, Ilam, Kerman, Kermanshah, Khuzestan, Kohkiluyeh and Buyer Ahmad, Kurdistan, Lorestan, Markazi, Mazandaran, North Khorasan, Qazvin, Qom, Razavi Khorasan, Semnan, Sistan and Baluchestan, South Khorasan, Tehran, West Azerbaijan, Yazd,

Zanjan (31 provinces)
Ethnic/Racial Groups: Persian, Azeri, Gilaki/Mazandarani, Kurd, Arab, Lur, Balochi, Turkmen
Religion: Islam (primarily Shiite and Sunni)
Languages: Persian (Farsi), Turkic, Kurdish, various local dialects
Currency: Iranian rial
Current President: Mahmud Ahmadinejad
Cities (capital, largest, or with at least a million people):
Tehran: • located in northern Iran • city in the Tehran province • capital of Iran • most populated city in Iran (7,190,000 people) • on the edge of the Elburz Mountains
Mashhad: • located in northeastern Iran • city in the Khorasan province • located in a valley in the Elburz Mountains
Esfahan: • located in central Iran • city in the Esfahan province • on the edge of the Zagros Mountains
Tabriz: • located in northwestern Iran • city in the Azarbayjan-e Gharbi province • located in a valley in the Zagros Mountains
Shiraz: • located in southwestern Iran • city in the Fars province • located in a valley in the Zagros Mountains
Karaj: • located in northern Iran • city in the Tehran province
Qom: • located in northern Iran • city in the Qom province • located in the Dasht-e Kavir on the edge of the Zagros Mountains

ENVIRONMENTAL/ECONOMIC:
Climate: arid to semi-arid; subtropical along the Caspian coast
Natural Resources: petroleum, natural gas, coal, chromium, copper
Agricultural Products: wheat, rice, other grains, sugar beets, dairy products
Major Exports: petroleum, carpets, fruits and nuts, iron and steel, chemicals
Natural Hazards: droughts, floods, dust storms, sandstorms, earthquakes

IRAQ
Country Name: Republic of Iraq
Continent: Asia
Area: 168,754 sq mi / 437,072 sq km
Population: 32,665,000
Population Density: 194 people per sq mi / 75 people per sq km
Capital: Baghdad

PHYSICAL:
Highest Point: unnamed peak 11,847 ft / 3,611 m
Lowest Point: Persian Gulf 0 ft / 0 m (sea level)
Mountain Ranges:
Zagros: • located in northeastern Iraq • extend into Turkey and Iran

Deserts:
  Syrian: •covers most of western Iraq •extends into Syria, Jordan, and Saudi Arabia
Gulfs:
  Persian: •forms the southern coast of Iraq •feeds into the Gulf of Oman
Lakes:
  Razzaza: •located in central Iraq •on the edge of the Syrian Desert
  Tharthar: •located in central Iraq •on the edge of the Syrian Desert
Rivers:
  Euphrates: •tributary of the Shatt al Arab •extends into Syria and Turkey
  Shatt al Arab: •has its mouth in the Persian Gulf •forms part of the Iraq-Iran border
  Tigris: •tributary of the Shatt al Arab •extends into Turkey

POLITICAL:
Independence: October 3, 1932 (from British administration)
Bordering Countries: Iran, Saudi Arabia, Syria, Turkey, Kuwait, Jordan (6)
Regions:
  Kurdistan: •region in northern Iraq •located in the Zagros Mountains •extends into Turkey and Iran
  Mesopotamia: •ancient region that covers an area from northwestern to southeastern Iraq •southwest of the Zagros Mountains
Administrative Divisions: Al Anbar, Al Basrah, Al Muthanna, Al Qadisiyah, An Najaf, Arbil, As Sulaymaniyah, Babil, Baghdad, Dahuk, Dhi Qar, Diyala, Karbala, Kirkuk, Maysan, Ninawa, Salah ad Din, Wasit (18 governorates) •Kurdistan Regional Government
Ethnic/Racial Groups: Arab, Kurdish

Religion: Islam (primarily Shiite and Sunni)
Languages: Arabic, Kurdish, Assyrian, Armenian
Currency: Iraqi dinar
Current President: Jalal Talabani
Cities (capital, largest, or with at least a million people):
  Baghdad: •located in central Iraq •city in the Baghdad governorate •capital of Iraq •most populated city in Iraq (5,751,000 people) •chief port on the Tigris River
  Mawsil (Mosul): •located in northern Iraq •city in the Ninawa governorate •a major port on the Tigris River
  Basrah (Basra): •located in southeastern Iraq •city in the Al Basrah governorate •chief port on the Shatt al Arab

ENVIRONMENTAL/ECONOMIC:
Climate: desert; mild to cool winters/hot summers; cold winters in northern mountains
Natural Resources: petroleum, natural gas, phosphates, sulfur
Agricultural Products: wheat, barley, rice, vegetables, cattle
Major Exports: crude oil
Natural Hazards: dust storms, sandstorms, floods

IRELAND
Country Name: Ireland
Continent: Europe
Area: 27,133 sq mi / 70,273 sq km
Population: 4,485,000
Population Density: 165 people per sq mi / 64 people per sq km
Capital: Dublin

PHYSICAL:
Highest Point: Carrantoohil 3,415 ft / 1,041 m
Lowest Point: Atlantic Ocean 0 ft / 0 m (sea level)
Mountain Ranges:
  Wicklow: •located in eastern Ireland
Oceans:
  Atlantic: •forms most of the western coast of Ireland
Seas:
  Celtic: •forms most of the southern coast of Ireland •feeds into the Atlantic Ocean
  Irish: •forms part of the eastern coast of Ireland •feeds into the Celtic Sea and the Atlantic Ocean
Bays:
  Clew: •forms part of western coast of Ireland •feeds into the Atlantic Ocean
  Dingle: •forms part of the southwestern coast of Ireland •feeds into the Atlantic Ocean
  Donegal: •forms part of the northern coast of Ireland •feeds into the Atlantic Ocean

Dundalk: •forms part of the northeastern coast of Ireland •feeds into the Irish Sea
Galway: •forms part of the western coast of Ireland •feeds into the Atlantic Ocean
Channels:
  Saint George's: •forms part of the southeastern coast of Ireland •separates Ireland from the United Kingdom •connects the Irish and Celtic Seas
Lakes:
  Lough Corrib: •located in western Ireland
  Lough Derg: •located in central Ireland •fed by the Shannon River
  Lough Foyle: •located in northeastern Ireland •on the Ireland-United Kingdom border
Rivers:
  Shannon: •has its mouth in the Atlantic Ocean •has its source in northern Ireland
Regions:
  Connemara: •lowland region in western Ireland •north of the Atlantic Ocean and Galway Bay •west of Lough Corrib

POLITICAL:

Independence: December 6, 1921 (from the United Kingdom)
Bordering Country: United Kingdom (1)
Regions:
  Connacht: •region in western Ireland •east of the Atlantic Ocean
  Leinster: •located in eastern Ireland •includes the Wicklow Mountains •west of the Irish Sea
  Munster: •located in southern Ireland •north of the Celtic Sea •east of the Atlantic Ocean

Ulster: • located in northern Ireland • east of the Atlantic Ocean • extends into the United Kingdom
Administrative Divisions: Carlow, Cavan, Clare, Cork, Din Laoghaire-Rathdown, Donegal, Fingal, Galway, Kerry, Kildare, Kilkenny, Laois, Leitrim, Limerick, Longford, Louth, Mayo, Meath, Monaghan, North Tipperary, Offaly, Roscommon, Sligo, South Dublin, South Tipperary, Waterford, Westmeath, Wexford, Wicklow (30 counties) • Cork, Dublin, Galway, Limerick, Waterford (5 cities)
Ethnic/Racial Groups: Celtic, English
Religion: Christianity (primarily Roman Catholic)
Languages: English, Irish (Gaelic)
Currency: euro
Current President: Mary McAleese
Cities (capital, largest, or with at least a million people):
Dublin: • located in eastern Ireland • city in Dublin county • capital of Ireland • most populated city in Ireland (1,084,000 people) • chief port on the Irish Sea

ENVIRONMENTAL/ECONOMIC:
Climate: temperate maritime; moderated by the North Atlantic Current; mild winters/cool summers; humid
Natural Resources: natural gas, peat, copper, lead, zinc
Agricultural Products: turnips, barley, potatoes, sugar beets, beef
Major Exports: machinery and equipment, computers, chemicals, pharmaceuticals, live animals
Natural Hazards: floods, thunderstorms

ISRAEL
Country Name: State of Israel
Continent: Asia
Area: 8,550 sq mi / 22,145 sq km
Population: 7,856,000
Population Density: 919 people per sq mi / 355 people per sq km
Capital: Jerusalem

PHYSICAL:
Highest Point: Har Meron 3,963 ft / 1,208 m
Lowest Point: Dead Sea 1,385 ft / 422 m below sea level
Depressions:
Dead Sea: • located in eastern Israel • on the Israel-Jordan border • lowest point on land in the world
Deserts:
Negev: • covers most of southern Israel • extends into Egypt
Seas:
Mediterranean: • forms the western coast of Israel • feeds into the Atlantic Ocean

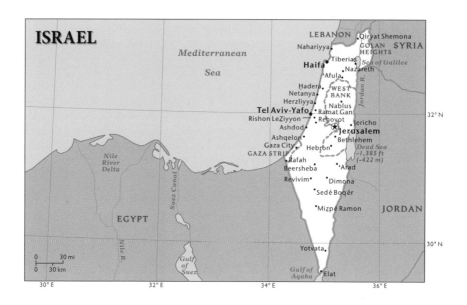

Gulfs:
  Gulf of Aqaba: •forms the southern coast of Israel •feeds into the Red Sea
Lakes:
  Sea of Galilee: •located in northeastern Israel
  Dead Sea: •located on the Israel-Jordan border •lowest point on land in the world
Rivers:
  Jordan: •has its mouth in the Dead Sea •forms part of the Israel-Jordan border

POLITICAL:
Independence: May 14, 1948 (from British administration)
Bordering Countries: Egypt, Jordan, Lebanon, Syria (4)
Regions:
  Golan Heights: •located in northeastern Israel •northeast of the Sea of Galilee
Administrative Divisions: Central, Haifa, Jerusalem, Northern, Southern, Tel Aviv (6 districts)
Palestinian Areas: Gaza Strip, West Bank (2)
Ethnic/Racial Groups: Jewish, Arab
Religions: Judaism, Islam, Christianity
Languages: Hebrew, Arabic, English
Currency: new Israeli shekel
Current Prime Minister: Binyamin Netanyahu
Cities (capital, largest, or with at least a million people):
  Tel Aviv-Yafo: •located in western Israel •city in the Tel Aviv district •most populated city in Israel (3,272,000 people) •a major port on the Mediterranean Sea
  Jerusalem: •located in central Israel •city in the Jerusalem district •capital of Israel (768,000 people)

ENVIRONMENTAL/ECONOMIC:
Climate: temperate to desert
Natural Resources: timber, potash, copper ore, natural gas, phosphate rock
Agricultural Products: citrus, vegetables, cotton, beef
Major Exports: machinery and equipment, software, cut diamonds, agricultural products, chemicals
Natural Hazards: sandstorms, droughts, earthquakes

ITALY

Country Name: Italian Republic
Continent: Europe
Area: 116,345 sq mi / 301,333 sq km
Population: 60,769,000
Population Density: 522 people per sq mi / 202 people per sq km
Capital: Rome

PHYSICAL:

Highest Point: Mont Blanc de Courmayeur 15,477 ft / 4,748 m
Lowest Point: Mediterranean Sea 0 ft / 0 m (sea level)
Mountain Ranges:
   Alps: •cover much of northern Italy •extend into France, Switzerland, Austria, and Slovenia
   Apennines: •cover an area from northwestern to southeastern Italy •extend into San Marino
   Dolomites: •located in northeastern Italy

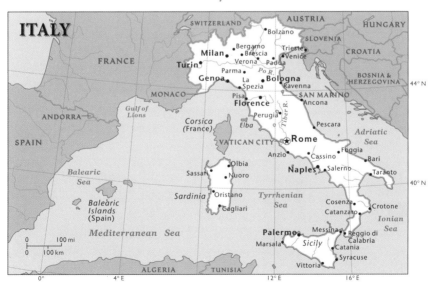

Peaks (minimum elevation 8,000 ft / 2,400 m):
    Dufourspitze: •located in the Alps •on the Italy-Switzerland border
    Mont Blanc de Coumayeur: •located in the Alps •on the Italy-France border
        •highest point in Italy
    Mount Etna: •located on the island of Sicily

Passes:
    Brenner Pass: •located in the Alps •connects Italy and Austria
    Great Saint Bernard: •located in the Alps •connects Italy and Switzerland

Valleys:
    Po: •covers much of northern Italy •between the Alps and the Apennines •eastern end borders the Gulf of Venice and the Adriatic Sea

Seas:
    Adriatic: •forms part of the eastern coast of Italy •feeds into the Ionian Sea
    Ionian: •forms part of the southeastern coast of Italy •feeds into the Mediterranean Sea
    Ligurian: •forms part of the northwestern coast of Italy •feeds into the Mediterranean Sea
    Mediterranean: •forms most of the western coast of Sardinia •forms most of the southern and eastern coasts of Sicily •feeds into the Atlantic Ocean
    Tyrrhenian: •forms most of the inner coast of Italy •feeds into the Mediterranean Sea

Gulfs:
    Gulf of Gaeta: •forms part of the western coast of Italy •feeds into the Tyrrhenian Sea
    Gulf of Genova: •forms part of the northwestern coast of Italy •feeds into the Ligurian Sea
    Gulf of Taranto: •forms part of the southeastern coast of Italy •feeds into the Ionian Sea
    Gulf of Trieste: •forms part of the northeastern coast of Italy •feeds into the Gulf of Venice
    Gulf of Venice: •forms part of the northeastern coast of Italy •feeds into the Adriatic Sea

Straits:
    Strait of Bonifacio: •forms part of the northern coast of Sardinia •separates the island of Sardinia from the French island of Corsica •connects the Mediterranean and Tyrrhenian Seas
    Strait of Messina: •forms part of the southern border of Italy •separates the island of Sicily from mainland Italy •connects the Tyrrhenian and Mediterranean Seas
    Strait of Otranto: •forms part of the southeastern coast of Italy •separates Italy from Albania •connects the Adriatic and Ionian Seas
    Strait of Sicily: •forms part of the western coast of Sicily •separates the island of Sicily from Tunisia

Channels:
    Malta: •forms part of the southern coast of Italy •separates Italy from Malta

Lakes:
    Lake Como: •located in northern Italy •in a valley of the Alps
    Lake Garda: •located in northern Italy •on the edge of the Alps
    Lake Maggiore: •located in northwestern Italy •in a valley of the Alps •extends into Switzerland

Rivers:
    Adige: •has its mouth in the Gulf of Venice •has its source in the Alps
    Po: •has its mouth in the Adriatic Sea •has its source in the Alps
    Tiber: •has its mouth in the Tyrrhenian Sea •has its source in the Appenines
Deltas:
    Mouths of the Po: •mouth of the Po River •feeds into the Adriatic Sea
Islands:
    Sardinia: •island off the western coast of mainland Italy •bordered by the Mediterranean Sea, Tyrrhenian Sea, and Strait of Bonifacio
    Sicily: •island off the southern coast of mainland Italy •bordered by the Mediterranean Sea, Tyrrhenian Sea, Strait of Sicily, Strait of Messina, and Malta Channel

POLITICAL:
Independence: March 17, 1861 (kingdom of Italy proclaimed; Italy was not finally unified until 1870)
Former Name: Kingdom of Italy
Bordering Countries: Switzerland, France, Austria, Slovenia, San Marino, Vatican City (6)
Administrative Divisions: Abruzzi, Apulia, Basilicata, Calabria, Campania, Emilia-Romagna, Friuli-Venezia Giulia, Latium, Liguria, Lombardy, Marches, Molise, Piedmont, Sardinia, Sicily, Tuscany, Trentino-Alto Adige, Umbria, Valle d'Aosta, Veneto (20 regions)
Ethnic/Racial Groups: predominantly Italian
Religion: Christianity (predominantly Roman Catholic)
Languages: Italian, German, French, Slovene
Currency: euro
Current President: Giorgio Napolitano
Cities (capital, largest, or with at least a million people):
    Rome: •located in central Italy •city in the Latium region •capital of Italy •most populated city in Italy (3,357,000 people) •chief port on the Tiber River
    Milan: •located in northern Italy •city in the Lombardy region
    Turin: •located in northern Italy •city in the Piedmont region

ENVIRONMENTAL/ECONOMIC:
Climate: Mediterranean; alpine in the far north; hot and dry in the south
Natural Resources: coal, mercury, zinc, potash, marble
Agricultural Products: fruits, vegetables, grapes, potatoes, beef
Major Exports: engineering products, textiles and clothing, production machinery, motor vehicles, transport equipment
Natural Hazards: landslides, mudflows, avalanches, earthquakes, volcanoes, floods, land subsidence in Venice

JAMAICA

Country Name: Jamaica
Continent: North America
Area: 4,244 sq mi / 10,991 sq km
Population: 2,709,000
Population Density: 638 people per sq mi / 247 people per sq km
Capital: Kingston

PHYSICAL:
Highest Point: Blue Mountain Peak 7,402 ft / 2,256 m
Lowest Point: Caribbean Sea 0 ft / 0 m (sea level)
Seas:
   Caribbean: •surrounds Jamaica •feeds into the Atlantic Ocean
Islands:
   Jamaica: •main island in Jamaica •surrounded by the Caribbean Sea

POLITICAL:
Independence: August 6, 1962 (from the United Kingdom)

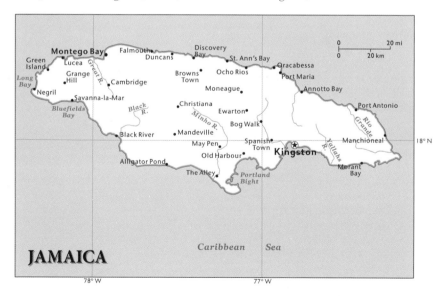

Administrative Divisions: Clarendon, Hanover, Kingston, Manchester, Portland, Saint Andrew, Saint Ann, Saint Catherine, Saint Elizabeth, Saint James, Saint Mary, Saint Thomas, Trelawny, Westmoreland (14 parishes)
Ethnic/Racial Groups: black, mixed, East Indian
Religions: Christianity (Protestant, Roman Catholic), other spiritual beliefs
Languages: English, patois English
Currency: Jamaican dollar
Current Prime Minister: Andrew Holness
Cities (capital, largest, or with at least a million people):
  Kingston: •located in eastern Jamaica •city in the Kingston parish •capital of Jamaica •most populated city in Jamaica (580,000 people) •chief port on the Caribbean Sea

ENVIRONMENTAL/ECONOMIC:
Climate: tropical; hot and humid with temperate interior
Natural Resources: bauxite, gypsum, limestone
Agricultural Products: sugarcane, bananas, coffee, citrus, poultry
Major Exports: alumina, bauxite, sugar, bananas, rum
Natural Hazards: hurricanes

JAPAN
Country Name: Japan
Continent: Asia
Area: 145,902 sq mi / 377,887 sq km
Population: 128,100,000
Population Density: 878 people per sq mi / 339 people per sq km
Capital: Tokyo

PHYSICAL:
Highest Point: Mount Fuji 12,388 ft / 3,776 m
Lowest Point: Hachiro-gata 13 ft / 4 m below sea level
Peaks (minimum elevation of 8,000 ft / 2,400 m):
  Mount Fuji: •located on the island of Honshu •highest point in Japan
Oceans:
  Pacific: •forms most of the eastern and southern coasts of Japan
Seas:
  East China: •forms most of the western coast of the island of Kyushu •feeds into the Pacific Ocean
  Inland: •forms part of the inner coast of Japan •feeds into the Pacific Ocean
  Sea of Japan: •forms most of the western coast of Japan •feeds into the Sea of Okhotsk, Pacific Ocean, and East China Sea
  Sea of Okhotsk: •forms most of the northern coast of the island of Hokkaido •feeds into the Pacific Ocean

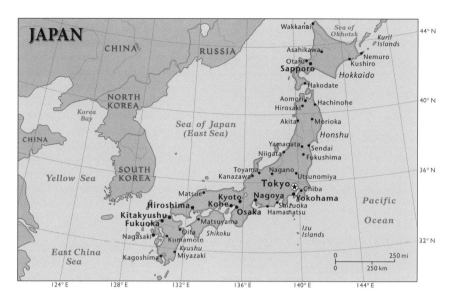

Bays:
Toyama: •forms part of the northern coast of Japan •feeds into the Sea of Japan
Straits:
Korea: •forms part of the southwestern coast of Japan •separates Japan from South Korea •connects the East China Sea and Sea of Japan
La Perouse: •forms part of the northwestern coast of Hokkaido •separates the island of Hokkaido from Russia •connects the Sea of Japan and Sea of Okhotsk
Tsugaru: •forms part of the northern coast of Japan •separates the island of Hokkaido from the island of Honshu •connects the Sea of Japan and Pacific Ocean
Lakes:
Lake Biwa: •located on the island of Honshu
Islands:
Hokkaido: •northernmost of the four main islands of Japan •bordered by the Pacific Ocean, Sea of Okhotsk, Sea of Japan, La Perouse Strait, and Tsugaru Strait
Honshu: •largest of the four main islands of Japan •bordered by the Pacific Ocean, Sea of Japan, Inland Sea, Toyama Bay, Tsugaru Strait, and Korea Strait
Kyushu: •southernmost of the four main islands of Japan •bordered by the Pacific Ocean, East China Sea, and Korea Strait
Ryukyu: •island group located south of Kyushu •bordered by the East China Sea and the Pacific Ocean
Shikoku: •smallest of the four main islands of Japan •bordered by the Pacific Ocean and Inland Sea

POLITICAL:
Independence: 660 B.C. (traditional founding by Emperor Jimmu)
Administrative Divisions: Aichi, Akita, Aomori, Chiba, Ehime, Fukui, Fukuoka, Fukushima, Gifu, Gumma, Hiroshima, Hokkaido, Hyogo, Ibaraki, Ishikawa, Iwate, Kagawa, Kagoshima, Kanagawa, Kochi, Kumamoto, Kyoto, Mie, Miyagi, Miyazaki, Nagano, Nagasaki, Nara, Niigata, Oita, Okayama, Okinawa, Osaka, Saga, Saitama, Shiga, Shimane, Shizuoka, Tochigi, Tokushima, Tokyo, Tottori, Toyama, Wakayama, Yamagata, Yamaguchi, Yamanashi (47 prefectures)
Ethnic/Racial Groups: predominantly Japanese
Religions: Shinto, Buddhism
Language: predominantly Japanese
Currency: yen
Current Prime Minister: Yoshihiko Noda
Cities (capital, largest, or with at least a million people):
Tokyo-Yokohama: •located on the island of Honshu •cities in the Tokyo, Saitama, Kanagawa, and Chiba prefectures •Tokyo is the capital of Japan •most populated metropolitan area in the world (36,507,000 people) •Yokohama is a major port on the Pacific Ocean
Osaka-Kobe-Kyoto: •located on the island of Honshu •cities in the Osaka, Hyogo, Kyoto, and Nara prefectures •a major port on the Pacific Ocean and Lake Biwa
Nagoya: •located on the island of Honshu •city in the Aichi and Gifu prefectures •a major port on the Pacific Ocean
Sapporo: •located on the island of Hokkaido •city in the Hokkaido prefecture
Fukuoka: •located on the island of Kyushu •city in the Fukuoka prefecture
Hiroshima: •located on the island of Honshu •city in the Hiroshima prefecture •chief port on the Inland Sea
Kitakyushu: •located on the island of Kyushu •city in the Fukuoka prefecture
Sendai: •located on the island of Honshu •city in the Miyagi prefecture

ENVIRONMENTAL/ECONOMIC:
Climate: tropical in the south; cool temperate in the north
Natural Resources: negligible mineral resources, fish
Agricultural Products: rice, sugar beets, vegetables, fruits, pork
Major Exports: motor vehicles, semiconductors, office machinery, chemicals
Natural Hazards: volcanoes, earthquakes, tsunamis, typhoons

JORDAN
Country Name: Hashemite Kingdom of Jordan
Continent: Asia
Area: 34,495 sq mi / 89,342 sq km
Population: 6,632,000
Population Density: 192 people per sq mi / 74 people per sq km
Capital: Amman

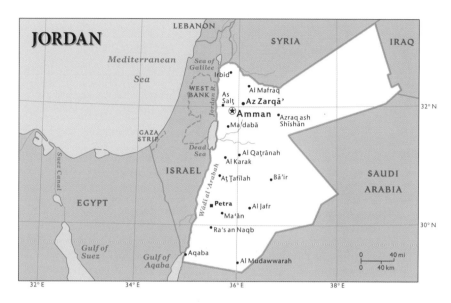

PHYSICAL:
Highest Point: Jabal Umm al Dami 6,083 ft / 1,854 m
Lowest Point: Dead Sea 1,385 ft / 422 m below sea level
Depressions:
  Dead Sea: •located in western Jordan •on the Jordan-Israel border •lowest point on land in the world
Deserts:
  Syrian: •covers much of central and eastern Jordan •extends into Syria, Iraq, and Saudi Arabia
Gulfs:
  Gulf of Aqaba: •forms the southern coast of Jordan •feeds into the Red Sea
Lakes:
  Dead Sea: •on the Jordan-Israel border •lowest point on land in the world
Rivers:
  Jordan: •has its mouth in the Dead Sea •forms part of the Jordan-Israel border

POLITICAL:
Independence: May 25, 1946 (from British administration)
Former Name: Transjordan
Bordering Countries: Saudi Arabia, Syria, Israel, Iraq (4)
Administrative Divisions: Ajlun, Al Aqabah, Al Balqa, Al Karak, Al Mafraq, Amman, At Tafilah, Az Zarqa, Irbid, Jarash, Maan, Madaba (12 governorates)
Ethnic/Racial Groups: predominantly Arab
Religions: Islam (primarily Sunni), Christianity

Languages: Arabic, English
Currency: Jordanian dinar
Current Prime Minister: Awn Khasawneh
Cities (capital, largest, or with at least a million people):
- Amman: •located in northwestern Jordan •city in the Amman governorate •capital of Jordan •most populated city in Jordan (1,088,000 people) •on the edge of the Syrian Desert

ENVIRONMENTAL/ECONOMIC:
Climate: mostly arid desert with rainy/dry seasons in the west
Natural Resources: phosphates, potash, shale oil
Agricultural Products: wheat, barley, citrus, tomatoes, sheep
Major Exports: phosphates, fertilizers, potash, agricultural products, manufactures
Natural Hazards: droughts, earthquakes

COUNTRIES

KAZAKHSTAN

Country Name: Republic of Kazakhstan
Continent: Asia
Area: 1,049,155 sq mi / 2,717,300 sq km
Population: 16,553,000
Population Density: 16 people per sq mi / 6 people per sq km
Capital: Astana

PHYSICAL:

Highest Point: Khan Tangiri 22,949 ft / 6,995 m
Lowest Point: Vpadina Kaundy (in the Caspian Depression) 433 ft / 132 m below sea level
Mountain Ranges:
  Altay: •located in eastern Kazakhstan •extend into Russia and China
  Tian Shan: •located in southeastern Kazakhstan •extend into Kyrgyzstan and China
Peaks (minimum elevation of 8,000 ft / 2,400 m):
  Khan Tangiri: •located in the Tian Shan •on Kazakhstan's border with China and Kyrgyzstan •highest point in Kazakhstan

Depressions:
　Caspian Depression: ●located in western Kazakhstan ●north of the Caspian Sea ●extends into Russia ●includes the lowest point in Kazakhstan
Plateau Regions:
　Kazakh Uplands: ●uplands in central Kazakhstan
　Ustyurt Plateau: ●basin in southwestern Kazakhstan ●extends into Uzbekistan
Deserts:
　Qizilqum: ●located in southern Kazakhstan ●extends into Uzbekistan
Lakes:
　Aral Sea: ●located in southwestern Kazakhstan ●on the Kazakhstan-Uzbekistan border
　Caspian Sea: ●forms the western coast of Kazakhstan ●also borders Russia and Turkmenistan ●largest lake in the world
　Lake Balkhash: ●located in eastern Kazakhstan
Rivers:
　Ertis (Irtysh): ●located in northeastern Kazakhstan ●extends into Russia and China
　Syr Darya: ●has its mouth in the Aral Sea ●extends into Uzbekistan

POLITICAL:
Independence: December 16, 1991 (from the Soviet Union)
Former Name: Kazakh Soviet Socialist Republic
Bordering Countries: Russia, Uzbekistan, China, Kyrgyzstan, Turkmenistan (5)
Administrative Divisions: Almaty, Aqmola, Aqtobe, Atyrau, Batys Qazaqstan, Mangghystau, Ongtustik Qazaqstan, Pavlodar, Qaraghandy, Qostanay, Qyzylorda, Shyghys Qazaqstan, Soltustik Qazaqstan, Zhambyl (14 provinces) ●Almaty, Astana, Bayqongyr (3 cities)
Ethnic/Racial Groups: Kazakh, Russian, Ukrainian, Uzbek
Religions: Islam, Christianity (primarily Russian Orthodox)
Languages: Kazakh, Russian
Currency: tenge
Current President: Nursultan Nazarbayev
Cities (capital, largest, or with at least a million people):
　Almaty: ●located in southeastern Kazakhstan ●city in the Almaty city division ●most populated city in Kazakhstan (1,383,000 people) ●on the edge of the Tian Shan
　Astana: ●located in northern Kazakhstan ●city in the Astana city division ●capital of Kazakhstan (650,000 people)

ENVIRONMENTAL/ECONOMIC:
Climate: continental; cold winters/hot summers; arid to semi-arid
Natural Resources: petroleum, natural gas, coal, iron ore, manganese
Agricultural Products: grain, cotton, livestock
Major Exports: oil and oil products, ferrous metals, chemicals, machinery, grain
Natural Hazards: earthquakes, mudslides

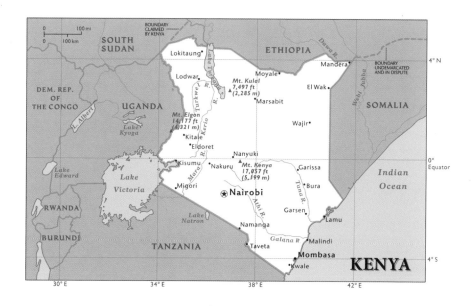

KENYA

Country Name: Republic of Kenya
Continent: Africa
Area: 224,081 sq mi / 580,367 sq km
Population: 41,610,000
Population Density: 186 people per sq mi / 72 people per sq km
Capital: Nairobi

PHYSICAL:
Highest Point: Mount Kenya 17,057 ft / 5,199 m
Lowest Point: Indian Ocean 0 ft / 0 m (sea level)
Mountain Ranges:
  Kenya Highlands: •covers much of central Kenya
Peaks (minimum elevation of 8,000 ft / 2,400 m):
  Mount Kenya: •located in the Kenya Highlands •highest point in Kenya
Valleys:
  Great Rift: •covers an area from northwestern to southwestern Kenya •cuts through the Kenya Highlands •extends into Ethiopia, Uganda, and Tanzania
Grasslands/Prairies:
  Serengeti Plain: •located in southwestern Kenya •extends into Tanzania
Deserts:
  Chalbi: •located in northern Kenya

Oceans:
  Indian: • forms the southern coast of Kenya
Lakes:
  Lake Turkana: • located in northwestern Kenya • in the Great Rift Valley • extends into Ethiopia
  Lake Victoria: • located in southwestern Kenya • extends into Uganda and Tanzania • largest lake in Africa • third largest lake in the world
Rivers:
  Tana: • has its mouth in the Indian Ocean • has its source in the Kenya Highlands

POLITICAL:
Independence: December 12, 1963 (from the United Kingdom)
Former Name: British East Africa
Bordering Countries: Uganda, Ethiopia, Tanzania, Somalia, South Sudan (5)
Administrative Divisions: Central, Coast, Eastern, North-Eastern, Nyanza, Rift Valley, Western (7 provinces) • Nairobi Area (1 area)
Ethnic/Racial Groups: Kikuyu, Luhya, Luo, Kalenjin, Kamba, Kisii, Meru
Religions: Christianity (Protestant, Roman Catholic), indigenous beliefs, Islam
Languages: English, Kiswahili, numerous indigenous languages
Currency: Kenyan shilling
Current President: Mwai Kibaki
Cities (capital, largest, or with at least a million people):
  Nairobi: • located in southern Kenya • city in the Nairobi Area • capital of Kenya • most populated city in Kenya (3,375,000 people) • on the edge of the Kenya Highlands

ENVIRONMENTAL/ECONOMIC:
Climate: tropical along the coast; arid in the interior
Natural Resources: gold, limestone, soda ash, salt, rubies
Agricultural Products: tea, coffee, corn, wheat, dairy products
Major Exports: tea, horticultural products, coffee, petroleum products, fish
Natural Hazards: droughts, floods

KIRIBATI
Country Name: Republic of Kiribati
Continent: Australia/Oceania
Area: 313 sq mi / 811 sq km
Population: 103,000
Population Density: 329 people per sq mi / 127 people per sq km
Capital: Tarawa

PHYSICAL:
Highest Point: unnamed location 266 ft / 81 m
Lowest Point: Pacific Ocean 0 ft / 0 m (sea level)
Oceans:
   Pacific: •forms the entire coast of Kiribati
Islands:
   Gilbert: •island group in western Kiribati •surrounded by the Pacific Ocean
   Line: •island group in eastern Kiribati •surrounded by the Pacific Ocean
   Phoenix: •island group in central Kiribati •surrounded by the Pacific Ocean

POLITICAL:
Independence: July 12, 1979 (from the United Kingdom)
Former Name: Gilbert Islands
Administrative Divisions: Gilbert Islands (includes Tarawa district), Line Islands, Phoenix Islands (3 units)
Ethnic/Racial Groups: predominantly Micronesian
Religion: Christianity (Roman Catholic, Protestant)
Languages: English, I-Kiribati
Currency: Australian dollar
Current President: Anote Tong
Cities (capital, largest, or with at least a million people):
Tarawa (district): •located in the Gilbert island chain •city in the Gilbert Islands unit •capital of Kiribati •most populated city in Kiribati (43,000 people) •chief port on the Pacific Ocean

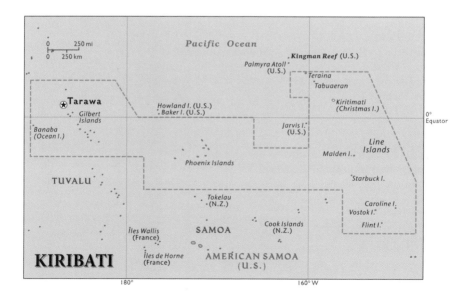

ENVIRONMENTAL/ECONOMIC:
Climate: tropical marine; hot and humid, moderated by trade winds
Natural Resources: phosphate
Agricultural Products: copra, taro, breadfruit, sweet potatoes, fish
Major Exports: copra, coconuts, seaweed, fish
Natural Hazards: typhoons, tornadoes

KOREA, NORTH

Country Name: Democratic People's Republic of Korea
Continent: Asia
Area: 46,540 sq mi / 120,538 sq km
Population: 24,457,000
Population Density: 526 people per sq mi / 203 people per sq km
Capital: Pyongyang

PHYSICAL:
Highest Point: Paektu-san 9,003 ft / 2,744 m
Lowest Point: Sea of Japan 0 ft / 0 m (sea level)
Peaks (minimum elevation of 8,000 ft / 2,400 m):
    Paektu-san: •located in northern North Korea •on the North Korea–China border •highest point in North Korea
Seas:
    Sea of Japan: •forms most of the eastern coast of North Korea •feeds into the East China Sea, Sea of Okhotsk, and Pacific Ocean

Yellow: •forms part of the southwestern coast of North Korea •feeds into the East China Sea

Bays:
Korea: •forms part of the western coast of North Korea •feeds into the Yellow Sea
Tonghan: •forms part of the eastern coast of North Korea •feeds into the Sea of Japan

POLITICAL:
Independence: August 15, 1945 (from Japan)
Bordering Countries: China, South Korea (2)
Administrative Divisions: Chagang, North Hamgyong, South Hamgyong, North Hwanghae, South Hwanghae, Kangwon, North Pyongan, South Pyongan, Ryanggang (9 provinces) •Nason, Pyongyang (2 municipalities)
Ethnic/Racial Groups: Korean
Religions: Buddhism, Confucianism
Language: Korean
Currency: North Korean won
Current President: Kim Jong Il
Cities (capital, largest, or with at least a million people):
Pyongyang: •located in southwestern North Korea •city in the Pyongyang special city •capital of North Korea •most populated city in North Korea (2,828,000 people)

ENVIRONMENTAL/ECONOMIC:
Climate: temperate with rainy summers
Natural Resources: coal, lead, tungsten, zinc, graphite
Agricultural Products: rice, corn, potatoes, soybeans, cattle
Major Exports: minerals, metallurgical products, manufactures, textiles
Natural Hazards: droughts, floods, typhoons

KOREA, SOUTH
Country Name: Republic of Korea
Continent: Asia
Area: 38,321 sq mi / 99,250 sq km
Population: 48,989,000
Population Density: 1,278 people per sq mi / 494 people per sq km
Capital: Seoul

PHYSICAL:
Highest Point: Hallasan 6,398 ft / 1,950 m
Lowest Point: Sea of Japan 0 ft / 0 m (sea level)
Seas:
East China: •forms the southern coast of the island of Jeju •feeds into the South China Sea and Pacific Ocean

Sea of Japan: •forms most of the eastern coast of South Korea •feeds into the East China Sea, Sea of Okhotsk, and Pacific Ocean •known as the East Sea to Koreans
Yellow: •forms most of the western coast of South Korea •feeds into the East China Sea
Straits:
Jeju: •forms part of the southwestern coast of South Korea •separates the island of Jeju from mainland South Korea
Korea: •forms part of the southern coast of South Korea •separates South Korea from Japan •connects the East China Sea and the Sea of Japan (East Sea)
Islands:
Jeju: •separated from mainland South Korea by the Jeju Strait •bordered by the Yellow Sea, the East China Sea, and the Jeju Strait

POLITICAL:
Independence: August 15, 1945 (from Japan)
Bordering Country: North Korea (1)
Administrative Divisions: North Chungcheong, South Chungcheong, Gangwon, Gyeonggi, North Gyeongsang, South Gyeongsang, Jeju, North Jeolla, South Jeolla (9 provinces) •Busan, Daegu, Daejeon, Gwangju, Incheon, Seoul, Ulsan (7 metropolitan cities)
Ethnic/Racial Groups: Korean
Religions: Christianity, Buddhism
Languages: Korean, English
Currency: South Korean won

Current President: Lee Myung-bak
Cities (capital, largest, or with at least a million people):
  Seoul: •located in northwestern South Korea •city in the Seoul metropolitan city division •capital of South Korea •most populated city in South Korea (9,778,000 people)
  Busan: •located in southeastern South Korea •city in the Busan metropolitan city division •chief port on the Korea Strait
  Daegu: •located in central South Korea •city in the Daegu metropolitan city division
  Incheon: •located in northwestern South Korea •city in the Incheon metropolitan city division •chief port on the Yellow Sea
  Daejeon: •located in western South Korea •city in the Daejeon metropolitan city division

ENVIRONMENTAL/ECONOMIC:
Climate: temperate, heavier rain in summer than in winter
Natural Resources: coal, tungsten, graphite, molybdenum, lead
Agricultural Products: rice, root crops, barley, vegetables, cattle
Major Exports: electronic products, machinery and equipment, motor vehicles, steel, ships, textiles
Natural Hazards: typhoons, minor earthquakes

KOSOVO
Country Name: Kosovo
Continent: Europe
Area: 4,203 sq mi / 10,887 sq km
Population: 2,284,000
Population Density: 544 people per sq mi / 210 people per sq km
Capital: Prishtina

PHYSICAL:
Highest Point: Deravica 8,714 ft / 2,656 m
Lowest Point: Beli Drim (located on the border with Albania) 974 ft / 297 m
Mountain Ranges:
  Sar Mountains: •situated along the southern border with Macedonia
  Kopaonik Mountains: •situated along the northeastern border with Serbia
Peaks (minimum elevation of 8,000 ft / 2,400 m):
  Deravica: •highest point in Kosovo
Plateau Regions:
  Kosovo Plain: •located in the east
  Dukagjin (Metohija) Plain: •located in the west

Rivers:
Sitnica River: •a tributary of the Ibër (Ibar) River
White Drin River: •flows through Kosovo and northern Albania •the headstream of the Drin River

POLITICAL:
Independence: February 17, 2008 (from Serbia)
Former Name: part of Yugoslavia
Bordering Countries: Macedonia, Albania, Serbia, Montenegro
Administrative Divisions: Decan, Dragash, Ferizaj, Fushe Kosove, Gjakova, Gjilan, Gllogovc/Drenas, Gracanica, Hani i Elezit, Istog, Junik, Kacanik, Kamenice/Dardana, Kline, Kllokot-Verboc, Leposaviq, Lipjan, Malisheva, Mamushe, Mitrovica, Novoberde, Obilic, Partesh, Peja, Podujevo, Prishtine Pristina, Prizren, Rahovec, Ranillug, Shterpca, Shtime, Skenderaj, Suhareka, Viti, Vushtrri, Zubin Potok, Zvecan (37 municipalities)
Ethnic/Racial Groups: Albanians, Serb, Bosniak, Gorani, Roma, Turk, Ashkali, Egyptian
Religion: Muslim, Serbian Orthodox, Roman Catholic
Languages: Albanian, Serbian, Bosnian, Turkish, Roma
Currency: euro
Current President: Atifete Jahjaga
Cities (capital, largest, or with at least a million people):
Prishtina: capital of Kosovo (600,000 people)

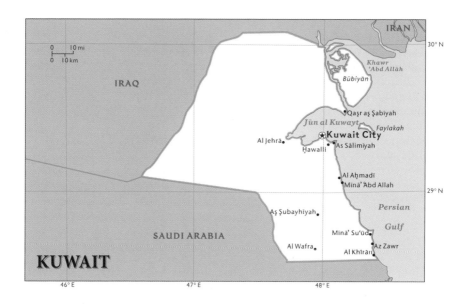

ENVIRONMENTAL/ECONOMIC:
Climate: moderate continental climate with warm summers

Natural Resources: nickel, lead zinc, magnesium, lignite, kaolin, chrome, bauxite
Agricultural: wheat, corn, berries, potatoes, peppers
Major Exports: mining and processed metal products, scrap metals, leather products, machinery, appliances
Natural Hazards: earthquakes

KUWAIT
Country Name: State of Kuwait
Continent: Asia
Area: 6,880 sq mi / 17,818 sq km
Population: 2,818,000
Population Density: 410 people per sq mi / 158 people per sq km
Capital: Kuwait City

PHYSICAL:
Highest Point: unnamed location 1,004 ft / 306 m
Lowest Point: Persian Gulf 0 ft / 0 m (sea level)
Gulfs:
  Persian Gulf: •forms most of the eastern coast of Kuwait •feeds into the Gulf

of Oman

POLITICAL:
Independence: June 19, 1961 (from the United Kingdom)
Bordering Countries: Iraq, Saudi Arabia (2)
Administrative Divisions: Al Ahmadi, Al 'Asimah, Al Farwaniyah, Al Jahra', Hawalli, Mubarak al Kabir (6 governorates)
Ethnic/Racial Groups: Arab, South Asian, Iranian
Religions: Islam (Sunni and Shiite), Christianity, Hinduism, Parsi
Languages: Arabic, English
Currency: Kuwaiti dinar
Current Amir: Sheikh Jabir al-Ahmad al-Jabir Al Sabah
Cities (capital, largest, or with at least a million people):
  Kuwait City: •located in eastern Kuwait •city in Al 'Asimah governorate •capital of Kuwait •most populated city in Kuwait (2,230,000 people) •chief port on the Persian Gulf

ENVIRONMENTAL/ECONOMIC:
Climate: dry desert with intensely hot summers/short and cool winters
Natural Resources: petroleum, fish, shrimp, natural gas
Agricultural Products: few crops, fish
Major Exports: oil and refined products, fertilizers
Natural Hazards: sudden rainstorms, sandstorms, dust storms

KYRGYZSTAN
Country Name: Kyrgyz Republic
Continent: Asia
Area: 77,182 sq mi / 199,900 sq km
Population: 5,600,000
Population Density: 73 people per sq mi / 28 people per sq km
Capital: Bishkek

PHYSICAL:
Highest Point: Jengish Chokusu (Pobedy Peak) 24,406 ft / 7,439 m
Lowest Point: Kara-Daryya 433 ft / 132 m
Mountain Ranges:
  Tian Shan: •cover an area from southwest to northeast Kyrgyzstan •extend into China
Peaks (minimum elevation of 8,000 ft / 2,400 m):
  Jengish Chokusu (Pobedy Peak): •located in the Tian Shan •on the Kyrgyzstan-China border

Lakes:
Ysyk Kol: •located in northeastern Kyrgyzstan •on the edge of the Tian Shan

POLITICAL:
Independence: August 31, 1991 (from the Soviet Union)
Former Name: Kirghiz Soviet Socialist Republic
Bordering Countries: Uzbekistan, Kazakhstan, Tajikistan, China (4)
Administrative Divisions: Batken, Chuy, Jalal-Abad, Naryn, Osh, Talas, Ysyk-Kol (7 provinces) •Bishkek (1 city)
Ethnic/Racial Groups: Kyrgyz, Russian, Uzbek, Ukrainian
Religions: Islam, Christianity (primarily Russian Orthodox)
Languages: Kyrgyz, Russian
Currency: Kyrgyzstani som
Current President: Almazbek Atambayev (president-elect, November 2011)
Cities (capital, largest, or with at least a million people):
Bishkek: •located in northern Kyrgyzstan •city in the Bishkek city division •capital of Kyrgyzstan •most populated city in Kyrgyzstan (854,000 people) •on the edge of the Tian Shan

ENVIRONMENTAL/ECONOMIC:
Climate: dry continental to polar in mountains; subtropical in southwest; temperate in northern foothills
Natural Resources: hydropower, gold and rare earth metals, coal, oil, natural gas
Agricultural Products: tobacco, cotton, potatoes, vegetables, sheep
Major Exports: cotton, wool, meat, tobacco, gold
Natural Hazards: floods, droughts

LAOS

Country Name: Lao People's Democratic Republic
Continent: Asia
Area: 91,429 sq mi / 236,800 sq km
Population: 6,259,000
Population Density: 68 people per sq mi / 26 people per sq km
Capital: Vientiane

PHYSICAL:
Highest Point: Phou Bia 9,245 ft / 2,818 m
Lowest Point: Mekong River 223 ft / 70 m
Mountain Ranges:
  Annam Cordillera: •covers most of eastern Laos •extends into Vietnam
Peaks (minimum elevation of 8,000 ft / 2,400 m):
  Phou Bia: •located on the Xiangkhoang Plateau •highest point in Laos
Plateau Regions:
  Plateau of Xiangkhoang: •mountainous plateau that covers much of northern Laos

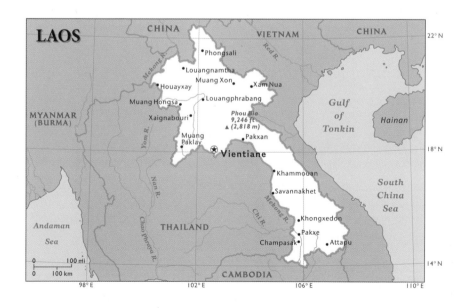

Rivers:
   Mekong: •forms part of Laos' border with Myanmar and Thailand •extends into China, Cambodia, and Vietnam

POLITICAL:
Independence: July 19, 1949 (from France)
Bordering Countries: Vietnam, Thailand, Cambodia, China, Myanmar (5)
Administrative Divisions: Attapu, Bokeo, Bolikhamxai, Champasak, Houaphan, Khammouan, Louangnamtha, Louangphrabang, Oudomxai, Phongsali, Salavan, Savannahkhet, Viangchan, Xaignabouli, Xekong, Xiangkhoang (16 provinces) •Viangchan (1 municipality) •Xiasomboun (1 special zone)
Ethnic/Racial Groups: Lao Loum, Lao Theung, Lao Soung
Religions: Buddhism, animism, other
Languages: Lao, French, English, various ethnic languages
Currency: kip
Current President: Choummali Saignason
Cities (capital, largest, or with at least a million people):
   Vientiane: • located in western Laos •city in the Viangchan municipality •capital of Laos •most populated city in Laos (799,000 people) •chief port on Mekong River

ENVIRONMENTAL/ECONOMIC:
Climate: tropical monsoon with rainy/dry seasons
Natural Resources: timber, hydropower, gypsum, tin, gold
Agricultural Products: sweet potatoes, vegetables, corn, coffee, water buffalo
Major Exports: wood products, garments, electricity, coffee, tin
Natural Hazards: floods, droughts

LATVIA
Country Name: Republic of Latvia
Continent: Europe
Area: 24,938 sq mi / 64,589 sq km
Population: 2,217,000
Population Density: 89 people per sq mi / 34 people per sq km
Capital: Riga

PHYSICAL:
Highest Point: Gaizina Kalns 1,024 ft / 312 m
Lowest Point: Baltic Sea 0 ft / 0 m (sea level)
Seas:
   Baltic: •forms part of the western coast of Latvia •feeds into the Kattegat
Gulfs:
   Gulf of Riga: • forms part of the northern and central coasts of Latvia •feeds into the Baltic Sea

Rivers:
Daugava: • has its mouth in the Gulf of Riga • extends into Belarus

POLITICAL:
Independence: August 21, 1991 (from the Soviet Union)
Former Name: Latvian Soviet Socialist Republic
Bordering Countries: Lithuania, Estonia, Russia, Belarus (4)
Administrative Divisions: Adazu, Aglonas, Aizkraukles, Aizputes, Aknistes, Alojas, Alsungas, Aluksnes, Amatas, Apes, Auces, Babites, Baldones, Baltinavas, Balvu, Bauskas, Beverinas, Brocenu, Burtnieku, Carnikavas, Cesu, Cesvaines, Ciblas, Dagdas, Daugavpils, Dobeles, Dundagas, Durbes, Engures, Erglu, Garkalnes, Grobinas, Gulbenes, Iecavas, Ikskiles, Ilukstes, Incukalna, Jaunjelgavas, Juanpiebalgas, Jaunpils, Jekabpils, Jelgavas, Kandavas, Karsavas, Keguma, Kekavas, Kocenu, Kokneses, Kraslavas, Krimuldas, Krustpils, Kuldigas, Lielvardes, Ligatnes, Limbazu, Livanu, Lubanas, Ludzas, Madonas, Malpils, Marupes, Mazsalacas, Nauksenu, Neretas, Nicas, Ogres, Olaines, Ozolnieku, Pargaujas, Pavilostas, Plavinu, Preilu, Priekules, Priekulu, Raunas, Rezeknes, Riebinu, Rojas, Ropazu, Rucavas, Rugaju, Rujienas, Rundales, Salacgrivas, Salas, Salaspils, Saldus, Saulkrastu, Sejas, Siguldas, Skriveru, Skrundas, Smiltenes, Stopinu, Strencu, Talsu, Tervetes, Tukuma, Vainodes, Valkas, Varaklanu, Varkavas, Vecpiebalgas, Vecumnieku, Ventspils, Viesites, Vilakas, Vilanu, Zilupes (109 municipalities)
Ethnic/Racial Groups: Latvian, Russian, Belarusian, Ukrainian, Polish
Religion: Christianity (Lutheran, Roman Catholic, Russian Orthodox)
Languages: Latvian, Lithuanian, Russian
Currency: Latvian lat

Current President: Andris Berzins
Cities (capital, largest, or with at least a million people):
 Riga: •located in central Latvia •city in the Riga municipality •capital of Latvia
  •most populated city in Latvia (711,000 people) •chief port on the
  Daugava River

ENVIRONMENTAL/ECONOMIC:
Climate: maritime, wet and moderate winters
Natural Resources: peat, limestone, dolomite, amber, hydropower
Agricultural Products: grain, sugar beets, potatoes, vegetables, beef
Major Exports: wood and wood products, machinery and equipment, metals, textiles
Natural Hazards: floods, snowstorms, thunderstorms

LEBANON
Country Name: Lebanese Republic
Continent: Asia
Area: 4,036 sq mi / 10,452 sq km
Population: 4,264,000
Population Density: 1,056 people per sq mi / 408 people per sq km
Capital: Beirut

PHYSICAL:
Highest Point: Qurnat as Sawda 10,131 ft / 3,088 m

Lowest Point: Mediterranean Sea 0 ft / 0 m (sea level)
Mountain Ranges:
  Anti-Lebanon : •located in eastern Lebanon •extend into Syria
  Lebanon: •located in central Lebanon
Peaks (minimum elevation of 8,000 ft / 2,400 m):
  Qurnat as Sawda: •located in the Lebanon Mountains •highest point in Lebanon
Valleys:
  Bekaa: •located between the Lebanon and Anti-Lebanon Mountains
Seas:
  Mediterranean: •forms the western coast of Lebanon •feeds into the Atlantic Ocean
Rivers:
  Orontes: •has its source in the Anti-Lebanon Mountains •extends into Syria

POLITICAL:
Independence: November 22, 1943 (from French administration)
Bordering Countries: Syria, Israel (2)
Administrative Divisions: Beqaa, Beyrouth, Liban-Nord, Liban-Sud, Mont-Liban, Nabatiye (6 governorates)
Ethnic/Racial Groups: Arab, Armenian
Religions: Islam, Christianity
Languages: Arabic, French, English, Armenian
Currency: Lebanese pound
Current President: Michel Sulayman
Cities (capital, largest, or with at least a million people):
  Beirut: •located in western Lebanon •city in the Beyrouth governorate •capital of Lebanon •most populated city in Lebanon (1,909,000 people) •chief port on the Mediterranean Sea

ENVIRONMENTAL/ECONOMIC:
Climate: Mediterranean; mild to cool and wet winters/hot and dry summers, heavy winter snows in the mountains
Natural Resources: limestone, iron ore, salt, water-surplus, arable land
Agricultural Products: citrus, grapes, tomatoes, apples, sheep
Major Exports: foodstuffs and tobacco, textiles, chemicals, precious stones, metal products
Natural Hazards: dust storms, sandstorms

LESOTHO
Country Name: Kingdom of Lesotho
Continent: Africa
Area: 11,720 sq mi / 30,355 sq km
Population: 2,194,000
Population Density: 187 people per sq mi / 72 people per sq km
Capital: Maseru

PHYSICAL:
Highest Point: Thabana-Ntlenyana 11,424 ft / 3,482 m
Lowest Point: Orange/Makhaleng River junction 4,593 ft / 1,400 m
Mountain Ranges:
   Drakensberg: •cover most of the country •extend into South Africa
Peaks (minimum elevation of 8,000 ft / 2,400 m):
   Thabana-Ntlenyana: •located in the Drakensberg •highest point in Lesotho
Rivers:
   Orange: •has its source in the Drakensberg •extends into South Africa

POLITICAL:
Independence: October 4, 1966 (from the United Kingdom)
Former Name: Basutoland
Bordering Country: South Africa (1)
Administrative Divisions: Berea, Butha-Buthe, Leribe, Mafeteng, Maseru, Mohale's Hoek, Mokhotlong, Qacha's Nek, Quthing, Thaba-Tseka (10 districts)
Ethnic/Racial Groups: predominantly Sotho
Religions: Christianity, indigenous beliefs
Languages: Sesotho, English, Zulu, Xhosa
Currency: loti, South African rand
Current Prime Minister: Pakalitha Bethuel Mosisili
Cities (capital, largest, or with at least a million people):
   Maseru: • located in western Lesotho •city in the Maseru district •capital of Lesotho •most populated city in Lesotho (220,000 people) •on the edge of the Drakensberg

ENVIRONMENTAL/ECONOMIC:
Climate: temperate; cool to cold with dry winters/hot and wet summers
Natural Resources: water, agricultural and grazing lands, some diamonds and other minerals
Agricultural Products: corn, wheat, pulses, sorghum, livestock
Major Exports: manufactures, wool, mohair
Natural Hazards: droughts

LIBERIA
Country Name: Republic of Liberia
Continent: Africa
Area: 43,000 sq mi / 111,370 sq km
Population: 4,133,000
Population Density: 96 people per sq mi / 37 people per sq km
Capital: Monrovia

PHYSICAL:
Highest Point: Mount Wuteve 4,528 ft / 1,380 m
Lowest Point: Atlantic Ocean 0 ft / 0 m (sea level)
Oceans:
   Atlantic: •forms the southern coast of Liberia

POLITICAL:
Independence: July 26, 1847 (established as an independent republic)
Bordering Countries: Cote d'Ivoire, Guinea, Sierra Leone (3)

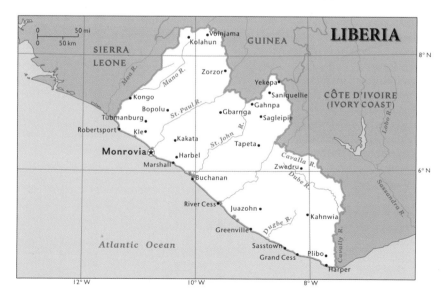

Administrative Divisions: Bomi, Bong, Gbarpolu, Grand Bassa, Grand Cape Mount, Grand Gedeh, Grand Kru, Lofa, Margibi, Maryland, Montserrado, Nimba, River Cess, River Gee, Sinoe (15 counties)
Ethnic/Racial Groups: Kpelle, Bassa, Dey, and other indigenous tribes
Religions: indigenous beliefs, Christianity, Islam
Languages: English, 20 ethnic languages
Currency: Liberian dollar
Current President: Ellen Johnson Sirleaf
Cities (capital, largest, or with at least a million people):
   Monrovia: • located in western Liberia • city in Montserrado county • capital of Liberia • most populated city in Liberia (882,000 people) • chief port on the Atlantic Ocean

ENVIRONMENTAL/ECONOMIC:
Climate: tropical; hot and humid with dry and moderate winters/wet summers
Natural Resources: iron ore, timber, diamonds, gold, hydropower
Agricultural Products: rubber, coffee, cacao, rice, sheep
Major Exports: rubber, timber, iron, diamonds, cacao
Natural Hazards: harmattan winds

LIBYA
Country Name: Great Socialist People's Libyan Arab Jamahiriya
Continent: Africa
Area: 679,362 sq mi / 1,759,540 sq km
Population: 6,423,000
Population Density: 9 people per sq mi / 4 people per sq km
Capital: Tripoli

PHYSICAL:
Highest Point: Bikku Bitti 7,438 ft / 2,267 m
Lowest Point: Sabkhat Ghuzayyil 154 ft / 47 m below sea level
Mountain Ranges:
   Tibesti: • located in southern Libya • extend into Chad
Plateau Regions:
   Libyan Plateau: • located in northeastern Libya • extends into Egypt • arid plateau north of the Sahara
Depressions:
   Sabkhat Ghuzayyil: • located in northern Libya • lowest point in Libya

Deserts:
   Libyan: • covers much of eastern Libya • extends into Egypt and Sudan
   Sahara: • covers most of central and southern Libya • extends into Tunisia, Algeria, Niger, Chad, Sudan, and Egypt • largest desert in the world

Seas:
Mediterranean: •forms part of the northern coast of Libya •feeds into the Atlantic Ocean
Gulfs:
Gulf of Sidra: •forms part of the northern coast of Libya •feeds into the Mediterranean Sea

POLITICAL:
Independence: December 24, 1951 (from Italy)
Bordering Countries: Egypt, Chad, Algeria, Tunisia, Sudan, Niger (6)
Regions:
Cyrenaica: •region in eastern Libya •includes part of the Sahara and the Libyan Desert
Fezzan: •region in southwestern Libya •includes part of the Sahara
Tripolitania: •region in northwestern Libya •south of the Mediterranean Sea •includes part of the Sahara
Administrative Divisions: Al Butnan, Al Jabal al Akhdar, Al Jabal al Gharbi, Al Jafarah, Al Jufrah, Al Kufrah, Al Marj, Al Marqab, Al Wahat, An Nuqat al Khams, Az Zawiyah, Banghazi, Darnah, Ghat, Misratah, Murzuq, Nalut, Sabha, Surt, Tarabulus, Wadi al Hayat, Wadi ash Shati (22 districts)
Ethnic/Racial Groups: predominantly Arab-Berber
Religion: Islam (predominantly Sunni)
Languages: Arabic, Italian, English
Currency: Libyan dinar
Transitional National Council Chairman: Mustafa Abd al-Jalil

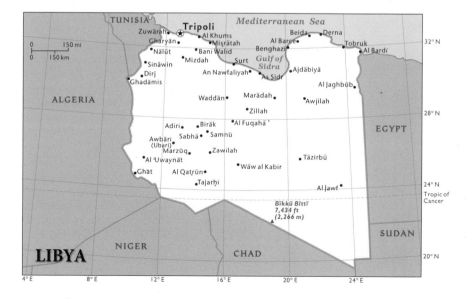

Cities (capital, largest, or with at least a million people):
  Tripoli: • located in northwestern Libya • city in the Tarabulas municipality • capital of Libya • most populated city in Libya (1,095,000 people) • chief port on the Mediterranean Sea

ENVIRONMENTAL/ECONOMIC:
Climate: Mediterranean along the coast; dry desert in the interior
Natural Resources: petroleum, natural gas, gypsum
Agricultural Products: wheat, barley, olives, dates, cattle
Major Exports: crude oil, refined petroleum products
Natural Disasters: ghibli winds, dust storms, sandstorms

LIECHTENSTEIN
Country Name: Principality of Liechtenstein
Continent: Europe
Area: 62 sq mi / 160 sq km
Population: 36,000
Population Density: 580 people per sq mi / 225 people per sq km
Capital: Vaduz

PHYSICAL:
Highest Point: Grauspitz 8,527 ft / 2,599 m
Lowest Point: Ruggeller Riet 1,411 ft / 430 m
Mountain Ranges:
  Alps: • cover most of Liechtenstein • extend into Switzerland and Austria

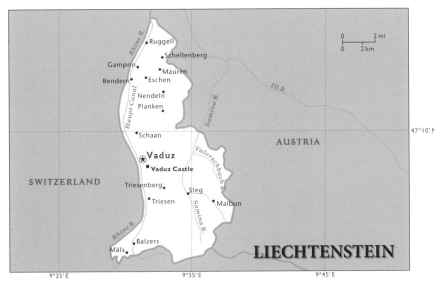

Peaks (minimum elevation of 8,000 ft / 2,400 m):
  Grauspitz: •located in the Alps •on the Liechtenstein-Switzerland border •highest point in Liechtenstein
Rivers:
  Rhine: • forms most of the Liechtenstein-Switzerland border •extends into Switzerland and Austria

POLITICAL:
Independence: January 23, 1719 (from the Holy Roman Empire)
Bordering Countries: Switzerland, Austria (2)
Administrative Divisions: Balzers, Eschen, Gamprin, Mauren, Planken, Ruggell, Schaan, Schellenberg, Triesen, Triesenberg, Vaduz (11 communes)
Ethnic/Racial Groups: Alemannic, Italian, Turkish
Religion: Christianity (Roman Catholic, Lutheran, Russian Orthodox, Protestant)
Languages: German, Alemannic dialect
Currency: Swiss franc
Current Prince: Hans Adam II
Cities (capital, largest, or with at least a million people):
  Vaduz: •located in western Liechtenstein •city in the Vaduz commune •capital of Liechtenstein •most populated city in Liechtenstein (5,000 people) •on the edge of the Alps •chief port on the Rhine River

ENVIRONMENTAL/ECONOMIC:
Climate: continental; cold winters/cool to moderate summers
Natural Resources: hydroelectric potential, arable land
Agricultural Products: wheat, barley, corn, potatoes, livestock
Major Exports: small specialty machinery, connectors for audio and video, parts for motor vehicles, dental products

LITHUANIA
Country Name: Republic of Lithuania
Continent: Europe
Area: 25,212 sq mi / 65,300 sq km
Population: 3,211,000
Population Density: 127 people per sq mi / 49 people per sq km
Capital: Vilnius

PHYSICAL:
Highest Point: Juozapines/Kalnas 958 ft / 292 m
Lowest Point: Baltic Sea 0 ft / 0 m (sea level)
Seas:
  Baltic: •forms the western coast of Lithuania •feeds into the Kattegat

Rivers:
   Nemunas: •has its mouth in the Baltic Sea •forms part of the Lithuania-Russia (Kaliningrad province) border •extends into Belarus

POLITICAL:
Independence: March 11, 1990 (from the Soviet Union)
Former Name: Lithuanian Soviet Socialist Republic
Bordering Countries: Belarus, Latvia, Russia (Kaliningrad province), Poland (4)
Administrative Divisions: Alytaus, Kauno, Klaipedos, Marijampoles, Panevezio, Siauliu, Taurages, Telsiu, Utenos, Vilniaus (10 counties)
Ethnic/Racial Groups: Lithuanian, Russian, Polish, Belarusian
Religion: Christianity (Roman Catholic, Lutheran, Russian Orthodox, Protestant)
Languages: Lithuanian, Polish, Russian
Currency: litas
Current President: Dalia Grybauskaite
Cities (capital, largest, or with at least a million people):
   Vilnius: •located in southeastern Lithuania •city in Vilniaus county •capital of Lithuania •most populated city in Lithuania (546,000 people)

ENVIRONMENTAL/ECONOMIC:
Climate: transitional; maritime to continental; wet and moderate
Natural Resources: peat, arable land
Agricultural Products: grain, potatoes, sugar beets, flax, beef
Major Exports: mineral products, textiles and clothing, machinery and equipment, chemicals

LUXEMBOURG

Country Name: Grand Duchy of Luxembourg
Continent: Europe
Area: 998 sq mi / 2,586 sq km
Population: 517,000
Population Density: 518 people per sq mi / 200 people per sq km
Capital: Luxembourg

PHYSICAL:
Highest Point: Buurgplaatz 1,834 ft / 559 m
Lowest Point: Moselle (Mosel) River 436 ft / 133 m
Rivers:
   Moselle (Mosel): •forms part of the Luxembourg-Germany border •extends into Germany and France

POLITICAL:
Independence: 1839 (from the Netherlands)
Bordering Countries: Belgium, Germany, France (3)
Regions:
   Ardennes: •forested plateau region in northern Luxembourg •extends into Belgium
Administrative Divisions: Diekirch, Grevenmacher, Luxembourg (3 districts)
Ethnic/Racial Groups: mixture of French and German
Religion: Christianity (Roman Catholic)
Languages: Luxembourgish, German, French
Currency: euro

Current Prime Minister: Jean-Claude Juncker
Cities (capital, largest, or with at least a million people):
  Luxembourg: •located in southern Luxembourg •city in the Luxembourg district •capital of Luxembourg •most populated city in Luxembourg (90,000 people)

ENVIRONMENTAL/ECONOMIC:
Climate: modified continental; mild winters/cool summers
Natural Resources: iron ore, arable land
Agricultural Products: barley, oats, potatoes, wheat, livestock
Major Exports: machinery and equipment, steel products, chemicals, rubber products, glass
Natural Hazards: floods, thunderstorms

COUNTRIES

MACEDONIA
Country Name: Republic of Macedonia
Continent: Europe
Area: 9,928 sq mi / 25,713 sq km
Population: 2,059,000
Population Density: 207 people per sq mi / 80 people per sq km
Capital: Skopje

PHYSICAL:
Highest Point: Golem Korab 9,068 ft / 2,764 m
Lowest Point: Vardar River 164 ft / 50 m
Mountain Ranges:
   Pindus: •cover much of western Macedonia •extend into Albania and Greece
Peaks (minimum elevation of 8,000 ft / 2,400 m):
   Golem Korab: •located in the Pindus Mountains •on the Macedonia-Albania border •highest point in Macedonia
Lakes:
   Lake Ohrid: •located in southwestern Macedonia •on the edge of the Pindus Mountains •on the Macedonia-Albania border
   Lake Prespa: •located in southwestern Macedonia •surrounded by the Pindus Mountains •on the border with Albania and Greece

POLITICAL:
Independence: September 8, 1991 (from Yugoslavia)
Former Names: People's Republic of Macedonia, Socialist Republic of Macedonia
Bordering Countries: Greece, Serbia, Albania, Bulgaria, Kosovo (5)
Regions:
   Macedonia: •region that stretches across southern Macedonia •includes part of the Pindus Mountains •extends into Greece and Bulgaria
Administrative Divisions: Aerodrom (Skopje), Aracinovo, Berovo, Bitola, Bogdanci, Bogovinje, Bosilovo, Brvenica, Butel (Skopje), Cair (Skopje), Caska, Centar (Skopje), Centar Zupa, Cesinovo, Cucer Sandevo, Debar, Debarca, Delcevo, Demir Hisar, Demir Kapija, Dojran, Dolneni, Dorce Petrov (Gjorce Petrov) (Skopje), Drugovo, Gazi Baba (Skopje), Gevgelija, Gostivar, Gradsko, Ilinden, Jegunovce, Karbinci, Karpos (Skopje), Kavadarci, Kicevo, Kisela Voda (Skopje), Kocani, Konce, Kratovo, Kriva Palanka, Krivogastani, Krusevo, Kumanovo, Lipkovo, Lozovo, Makedonska Kamenica, Makedonski Brod, Mavrovo i Rostusa, Mogila, Negotino, Novaci, Novo Selo, Ohrid, Oslomej, Pehcevo, Petrovec, Plasnica, Prilep, Probistip, Radovis, Rankovce, Resen, Rosoman, Saraj (Skopje), Sopiste, Staro Nagoricane, Stip, Struga, Strumica,

Studenicani, Suto Orizari (Skopje), Sveti Nikole, Tearce, Tetovo, Valandovo, Vasilevo, Veles, Vevcani, Vinica, Vranestica, Vrapciste, Zajas, Zelenikovo, Zelino, Zrnovci (84 municipalities)
Ethnic/Racial Groups: Macedonian, Albanian, Turkish
Religions: Christianity (primarily Macedonian Orthodox), Islam
Languages: Macedonian, Albanian, Turkish
Currency: Macedonian denar
Current President: Gjorge Ivanov
Cities (capital, largest, or with at least a million people):
  Skopje: •located in northern Macedonia •capital of Macedonia •most populated city in Macedonia (480,000 people)

ENVIRONMENTAL/ECONOMIC:
Climate: warm and dry summers/cold and snowy winters
Natural Resources: iron ore, copper, lead, zinc, chromite
Agricultural Products: rice, tobacco, wheat, corn, beef
Major Exports: food, beverages, tobacco, miscellaneous manufactures, iron and steel
Natural Hazards: earthquakes

MADAGASCAR
Country Name: Republic of Madagascar
Continent: Africa
Area: 226,658 sq mi / 587,041 sq km
Population: 21,315,000

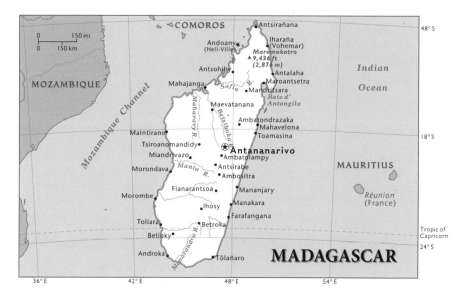

Population Density: 94 people per sq mi / 36 people per sq km
Capital: Antananarivo

PHYSICAL:
Highest Point: Maromokotro 9,436 ft / 2,876 m
Lowest Point: Indian Ocean 0 ft / 0 m (sea level)
Peaks (minimum elevation of 8,000 ft / 2,400 m):
   Maromokotro: •located in northeastern Madagascar •highest point in Madagascar
Oceans:
   Indian: •forms the eastern coast of Madagascar
Channels:
   Mozambique: •forms the western coast of Madagascar •separates the island of Madagascar from mainland Africa
Islands:
   Madagascar: •main island in Madagascar •bordered by the Indian Ocean and Mozambique Channel

POLITICAL:
Independence: June 26, 1960 (from France)
Former Name: Malagasy Republic
Administrative Divisions: Antananarivo, Antsiranana, Fianarantsoa, Mahajanga, Toamasina, Toliara (6 provinces)
Ethnic/Racial Groups: Malagasy, Cotiers, French, Indian, Chinese
Religions: indigenous beliefs, Christianity, Islam
Languages: French, Malagasy

Currency: Malagasy franc
Current President: Andry Rajoelina
Cities (capital, largest, or with at least a million people):
  Antananarivo: •located in central Madagascar •city in the Antananarivo province •capital of Madagascar •most populated city in Madagascar (1,816,000 people)

ENVIRONMENTAL/ECONOMIC:
Climate: tropical along the coast; temperate inland; arid in south
Natural Resources: graphite, chromite, coal, bauxite, salt
Agricultural Products: coffee, vanilla, sugarcane, cloves, livestock products
Major Exports: coffee, vanilla, shellfish, sugar, cotton cloth
Natural Hazards: cyclones, droughts, locust swarms

MALAWI
Country Name: Republic of Malawi
Continent: Africa
Area: 45,747 sq mi / 118,484 sq km
Population: 15,879,000
Population Density: 347 people per sq mi / 134 people per sq km
Capital: Lilongwe

PHYSICAL:
Highest Point: Sapitwa 9,849 ft / 3,002 m
Lowest Point: Shire River 121 ft / 37 m
Peaks (minimum elevation of 8,000 ft / 2,400 m):
  Sapitwa: •located in southeastern Malawi •highest point in Malawi
Valleys:
  Great Rift: •covers much of Malawi •extends into Zambia, Tanzania, and Mozambique
Lakes:
  Lake Malawi: •covers most of eastern Malawi •located in the Great Rift Valley •on the border with Tanzania and Mozambique •ninth largest lake in the world
Rivers:
  Shire: •has its source in Lake Malawi •extends into Mozambique

POLITICAL:
Independence: July 6, 1964 (from the United Kingdom)
Former Names: British Central African Protectorate, Nyasaland Protectorate, Nyasaland
Bordering Countries: Mozambique, Zambia, Tanzania (3)
Administrative Divisions: Balaka, Blantyre, Chikwawa, Chiradzulu, Chitipa, Dedza, Dowa, Karonga, Kasungu, Likoma, Lilongwe, Machinga (Kasupe), Mangochi, Mchinji, Mulanje, Mwanza, Mzimba, Neno, Nkhata Bay, Nkhotakota,

Nsanje, Ntcheu, Ntchisi, Phalombe, Rumphi, Salima, Thyolo, Zomba (28 districts)
Ethnic/Racial Groups: Chewa, Nyanja, Tumbuka, Yao, Lomwe, Sena, Tonga, Ngoni, Ngonde
Religions: Christianity (Protestant, Roman Catholic), Islam
Languages: English, Chichewa
Currency: Malawian kwacha
Current President: Bingu wa Mutharika
Cities (capital, largest, or with at least a million people):
Lilongwe: •located in western Malawi •city in the Lilongwe district •capital of Malawi •most populated city in Malawi (821,000 people)

ENVIRONMENTAL/ECONOMIC:
Climate: subtropical, rainy/dry seasons
Natural Resources: limestone, arable land, hydropower, uranium, coal
Agricultural Products: tobacco, sugarcane, cotton, tea, groundnuts
Major Exports: tobacco, tea, sugar, cotton, coffee

MALAYSIA
Country Name: Malaysia
Continent: Asia
Area: 127,355 sq mi / 329,847 sq km
Population: 28,885,000
Population Density: 227 people per sq mi / 88 people per sq km
Capital: Kuala Lumpur

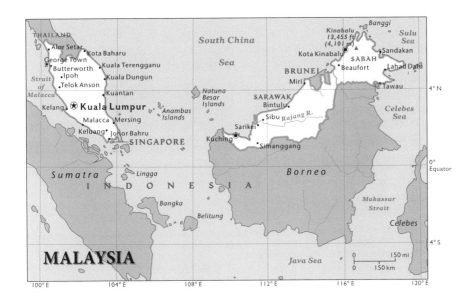

PHYSICAL:
Highest Point: Kinabalu 13,455 ft / 4,101 m
Lowest Point: South China Sea 0 ft / 0 m (sea level)
Peaks (minimum elevation of 8,000 ft / 2,400 m):
  Kinabalu: •located on the island of Borneo •highest point in Malaysia
Seas:
  Celebes: •forms part of the northeastern coast of the island of Borneo •feeds into the Philippine Sea
  South China: •forms the eastern coast of the Malay Peninsula and the northern coast of the island of Borneo •feeds into the Java and Philippine Seas
  Sulu: •forms part of the northeastern coast of the island of Borneo •feeds into the Celebes and South China Seas
Straits:
  Strait of Malacca: •forms the western coast of the Malay Peninsula •separates the Malay Peninsula from Indonesia •connects the Andaman and South China Seas
  Sigapore: •forms part of the southern coast of the Malay Peninsula •separates the Malay Peninsula from Indonesia •connects the Strait of Malacca and the South China Sea
Peninsulas:
  Malay: •bordered by the Strait of Malacca, South China Sea, and Singapore Strait •Thailand and Myanmar are located on the northern half
Islands:
  Borneo: •largest of the Greater Sunda Islands •bordered by the South China, Sulu, and Celebes Seas •parts of the island also belong to Brunei and Indonesia

POLITICAL:
Independence: August 31, 1957 (from the United Kingdom)
Former Name: Federation of Malaysia
Bordering Countries: Indonesia, Thailand, Brunei (3)
Regions:
   Sabah: •region on the island of Borneo •east of the South China Sea •west of the Sulu Sea
   Sarawak: •region on the island of Borneo •south of the South China Sea
Administrative Divisions: Johor, Kedah, Kelantan, Melaka, Negeri Sembilan, Pahang, Perak, Perlis, Pulau Pinang, Sabah, Sarawak, Selangor, Terengganu (13 states)
   •Wilayah Persekutuan (1 federal territory with 3 components: city of Kuala Lumpur, Labuan, and Putrajaya)
Ethnic/Racial Groups: Malay and other indigenous people, Chinese, East Indian
Religions: Islam, Buddhism, Daoism, Hinduism, Christianity, Sikh, Shamanism
Languages: Bahasa Melayu, English, Chinese dialects, other regional dialects and indigenous languages
Currency: ringgit
Current Sultan: Mizan Zainal Abidin
Cities (capital, largest, or with at least a million people):
   Kuala Lumpur: •located on the western side of the Malay Peninsula •city in Wilayah Persekutuan federal territory •capital of Malaysia •most populated city in Malaysia (1,494,000 people)

ENVIRONMENTAL/ECONOMIC:
Climate: tropical, annual monsoons
Natural Resources: tin, petroleum, timber, copper, iron ore
Agricultural Products: rubber, palm oil, cacao, rice
Major Exports: electronic equipment, petroleum and liquefied natural gas, wood and wood products, palm oil
Natural Hazards: floods, landslides, forest fires

MALDIVES
Country Name: Republic of Maldives
Continent: Asia
Area: 115 sq mi / 298 sq km (smallest in Asia)
Population: 325,000 (smallest in Asia)
Population Density: 2,826 people per sq mi / 1,091 people per sq km
Capital: Male

PHYSICAL:
Highest Point: unnamed location 8 ft / 2.4 m
Lowest Point: Indian Ocean 0 ft / 0 m (sea level)
Oceans:
   Indian: •forms most of the southern coast of the Maldives

Seas:
  Arabian: •forms the northwestern coast of the Maldives •feeds into the Indian Ocean
  Laccadive: •forms much of the eastern coast of the Maldives •feeds into the Indian Ocean
Islands:
  Maldives: •main island group of the Maldives •bordered by the Indian Ocean, Arabian Sea, and Laccadive Sea •consists of many coral reefs and atolls

POLITICAL:
Independence: July 26, 1965 (from the United Kingdom)
Administrative Divisions: Alifu, Baa, Dhaalu, Faafu, Gaafu Alifu, Gaafu Dhaalu, Gnaviyani, Haa Alifu, Haa Dhaalu, Kaafu, Laamu, Lhaviyani, Meemu, Noonu, Raa, Seenu, Shaviyani, Thaa, Vaavu (19 atolls) •Maale (1 first-order administrative division)
Ethnic/Racial Groups: Dravidian, Sinhalese, Arab
Religion: Islam (primarily Sunni)
Languages: Maldivian Dhivehi, English
Currency: rufiyaa
Current President: Mohamed "Anni" Nasheed
Cities (capital, largest, or with at least a million people):
  Male: •located in central Maldives •city in the Maale first-order administrative division •capital of the Maldives •most populated city in the Maldives (120,000 people) •chief port on the Indian Ocean

ENVIRONMENTAL/ECONOMIC:
Climate: tropical hot and humid with dry/rainy seasons

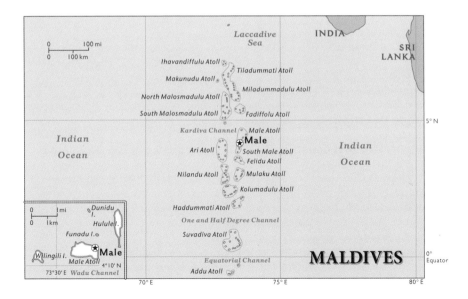

COUNTRIES A TO Z | PAGE 217

Natural Resources: fish
Agricultural Products: coconuts, corn, sweet potatoes, fish
Major Exports: fish, clothing
Natural Hazards: rising tides (flood lowland islands)

MALI

Country Name: Republic of Mali
Continent: Africa
Area: 478,841 sq mi / 1,240,192 sq km
Population: 15,394,000
Population Density: 32 people per sq mi / 12 people per sq km
Capital: Bamako

PHYSICAL:

Highest Point: Hombori Tondo 3,789 ft / 1,155 m
Lowest Point: Senegal River 75 ft / 23 m
Grasslands/Prairies:
  Sahel: •covers much of southern and central Mali •extends into Senegal, Guinea, Cote d'Ivoire, Burkina Faso, and Niger
Plateau Regions:
  Adrar des Iforas: •located in the Sahara in northeastern Mali •extends into Algeria
Swamps:
  Macina: •located in central Mali
Deserts:
  Sahara: •covers most of northern Mali and parts of central Mali •extends into Mauritania, Algeria, and Niger •largest desert in the world
Rivers:
  Niger: •extends into Guinea and Niger
  Senegal: •extends into Guinea •flows across the Mali-Senegal-Mauritania border

POLITICAL:

Independence: September 22, 1960 (from France)
Former Names: French Sudan, Sudanese Republic
Bordering Countries: Mauritania, Algeria, Burkina Faso, Guinea, Niger, Cote d'Ivoire, Senegal (7)
Administrative Divisions: Gao, Kayes, Kidal, Koulikoro, Mopti, Segou, Sikasso, Tombouctou (8 regions)
Ethnic/Racial Groups: Mande, Peul, Voltaic, Tuareg and Moor, Songhai
Religions: Islam, indigenous beliefs
Languages: French, Bambara, numerous African languages
Currency: Communaute Financiere Africaine franc
Current President: Amadou Toumani Toure
Cities (capital, largest, or with at least a million people):
  Bamako: •located in southern Mali •city in the Koulikoro region •capital of Mali •most populated city in Mali (1,628,000 people) •chief port on the Niger River

ENVIRONMENTAL/ECONOMIC:
Climate: subtropical to arid with hot summers/cool winters
Natural Resources: gold, phosphates, kaolin, salt, limestone
Agricultural Products: cotton, millet, rice, corn, cattle
Major Exports: cotton, gold, livestock
Natural Hazards: harmattan haze, droughts, floods

MALTA
Country Name: Republic of Malta
Continent: Europe
Area: 122 sq mi / 316 sq km
Population: 412,000
Population Density: 3,377 people per sq mi / 1,304 people per sq km
Capital: Valletta

PHYSICAL:
Highest Point: Ta'Dmejrek 830 ft / 253 m
Lowest Point: Mediterranean Sea 0 ft / 0 m (sea level)
Seas:
  Mediterranean: •forms most of the coast of Malta •feeds into the Atlantic Ocean
Channels:
  Malta: •forms the northern coast of Malta •separates Malta from Sicily
Islands:
  Gozo: •second largest island in Malta •borders the Mediterranean Sea and Malta Channel

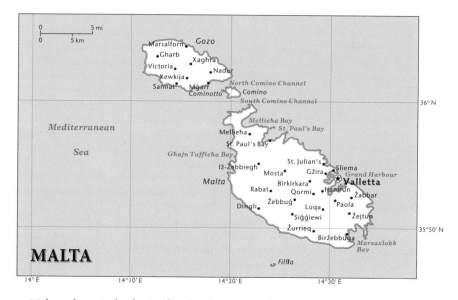

Malta: •largest island in Malta •borders the Mediterranean Sea and Malta Channel

POLITICAL:
Independence: September 21, 1964 (from the United Kingdom)
Ethnic/Racial Groups: predominantly Maltese
Religion: Christianity (primarily Roman Catholic)
Languages: Maltese, English
Currency: euro
Current President: George Abela
Cities (capital, largest, or with at least a million people):
  Valletta: •located on the island of Malta •capital of Malta •most populated city in Malta (199,000 people) •chief port on the Malta Channel

ENVIRONMENTAL/ECONOMIC:
Climate: Mediterranean with mild and rainy winters/hot and dry summers
Natural Resources: limestone, salt, arable land
Agricultural Products: potatoes, cauliflower, grapes, wheat, pork
Major Exports: machinery and transport equipment, manufactures
Natural Hazards: droughts

MARSHALL ISLANDS
Country Name: Republic of the Marshall Islands
Continent: Australia/Oceania
Area: 70 sq mi / 181 sq km

Population: 55,000
Population Density: 786 people per sq mi / 304 people per sq km
Capital: Majuro

PHYSICAL:
Highest Point: unnamed location 33 ft / 10 m
Lowest Point: Pacific Ocean 0 ft / 0 m (sea level)
Oceans:
  Pacific: •surrounds the Marshall Islands
Islands:
  Ralik Chain: •western chain of atolls and coral islands in the Marshall Islands
    •surrounded by the Pacific Ocean
  Ratak Chain: •eastern chain of atolls and coral islands in the Marshall Islands
    •surrounded by the Pacific Ocean

POLITICAL:
Independence: October 21, 1986 (from United States administration)
Former Name: Marshall Islands District (part of the Trust Territory of the Pacific Islands)
Administrative Divisions: Ailinginae, Ailinglaplap, Ailuk, Arno, Aur, Bikar, Bikini, Bokak, Ebon, Enewetak, Erikub, Jabat, Jaluit, Jemo, Kili, Kwajalein, Lae, Lib, Likiep, Majuro, Maloelap, Mejit, Mili, Namorik, Namu, Rongelap, Rongrik, Toke, Ujae, Ujelang, Utirik, Wotho, Wotje (33 municipalities)
Ethnic/Racial Groups: Micronesian
Religion: Christianity (primarily Protestant)
Languages: English, Marshallese, Japanese
Currency: U.S. dollar

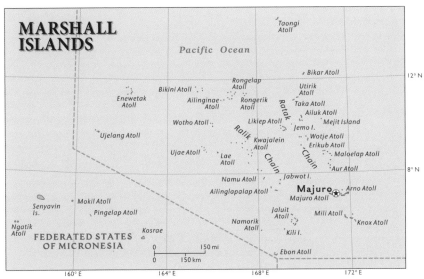

Current President: Jureland Zedkaia
Cities (capital, largest, or with at least a million people):
  Majuro: •located in the Ratak Chain •city in the Majuro municipality •capital of the Marshall Islands •most populated city in the Marshall Islands (30,000 people) •chief port on the Pacific Ocean

ENVIRONMENTAL/ECONOMIC:
Climate: tropical; hot and humid
Natural Resources: coconut products, marine products, deep seabed minerals
Agricultural Products: coconuts, tomatoes, melons, taro, pigs
Major Exports: copra cake, coconut oil, handicrafts, fish
Natural Hazards: typhoons

MAURITANIA
Country Name: Islamic Republic of Mauritania
Continent: Africa
Area: 397,955 sq mi / 1,030,700 sq km
Population: 3,542,000
Population Density: 9 people per sq mi / 3.5 people per sq km
Capital: Nouakchott

PHYSICAL:
Highest Point: Kediet ej Jill 3,002 ft / 915 m
Lowest Point: Sebkha de Ndrhamcha 10 ft / 5 m below sea level

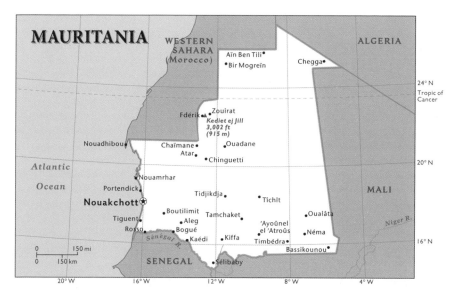

Deserts:
  Sahara: •covers most of the country •extends into Algeria, Mali, and Western Sahara (claimed by Morocco) •largest desert in the world
Oceans:
  Atlantic: •forms the western coast of Mauritania
Rivers:
  Senegal: •has its mouth in the Atlantic Ocean •forms all of the Mauritania-Senegal border •extends into Mali

POLITICAL:
Independence: November 28, 1960 (from France)
Bordering Countries: Mali, Senegal, Algeria, Western Sahara (claimed by Morocco) (4)
Administrative Divisions: Adrar, Assaba, Brakna, Dakhlet Nouadhibou, Gorgol, Guidimaka, Hodh Ech Chargui, Hodh El Gharbi, Inchiri, Tagant, Tiris Zemmour, Trarza (12 regions) •Nouakchott (1 capital district)
Ethnic/Racial Groups: mixed Maur/black, Maur, black
Religion: predominantly Islam
Languages: Hassaniya Arabic, Wolof, Pulaar, Soninke, French
Currency: ouguiya
Current President: Mohamed Ould Abdel Aziz
Cities (capital, largest, or with at least a million people):
  Nouakchott: •located in southwestern Mauritania •city in the Nouakchott capital district •capital of Mauritania •most populated city in Mauritania (709,000 people) •chief port on the Atlantic Ocean

ENVIRONMENTAL/ECONOMIC:
Climate: desert; hot and dry
Natural Resources: iron ore, gypsum, copper, phosphate, diamonds
Agricultural Products: dates, millet, sorghum, rice, cattle
Major Exports: iron ore, fish and fish products, gold
Natural Hazards: sirocco winds, droughts

MAURITIUS
Country Name: Republic of Mauritius
Continent: Africa
Area: 788 sq mi / 2,040 sq km
Population: 1,286,000
Population Density: 1,632 people per sq mi / 630 people per sq km (most dense in Africa)
Capital: Port Louis

PHYSICAL:
Highest Point: Piton de la Riviere Noire, 2,710 ft / 826 m
Lowest Point: Indian Ocean 0 ft / 0 m (sea level)

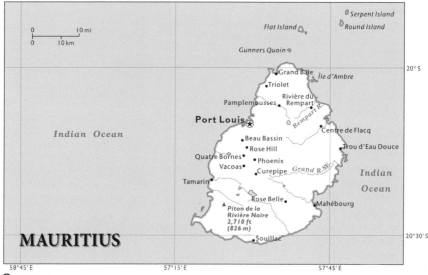

MAURITIUS

Oceans:
Indian: •surrounds Mauritius
Islands:
Mauritius: •main island in Mauritius •surrounded by the Indian Ocean

POLITICAL:
Independence: March 12, 1968 (from the United Kingdom)
Administrative Divisions: Black River, Flacq, Grand Port, Moka, Pamplemousses, Plaines Wilhems, Port Louis, Riviere du Rempart, Savanne (9 districts) •Agalega Islands, Cargados Carajos, Rodrigues (3 dependencies north and east of Mauritius)
Ethnic/Racial Groups: Indo-Mauritian, Creole, Sino-Mauritian, Franco-Mauritian
Religions: Hinduism, Christianity (primarily Roman Catholic), Islam
Languages: English, French, Creole, Hindi, Urdu, Hakka, Bhojpuri
Currency: Mauritian rupee
Current President: Anerood Jugnauth
Cities (capital, largest, or with at least a million people):
Port Louis: •located in northwestern Mauritius •city in the Port Louis district •capital of Mauritius •most populated city in Mauritius (149,000 people) •chief port on the Indian Ocean

ENVIRONMENTAL/ECONOMIC:
Climate: tropical, moderated by southeast trade winds; dry winters/hot and rainy summers
Natural Resources: arable land, fish
Agricultural Products: sugarcane, tea, corn, potatoes, bananas
Major Exports: clothing and textiles, sugar, cut flowers, molasses
Natural Hazards: cyclones, maritime hazards

MEXICO

Country Name: United Mexican States
Continent: North America
Area: 758,449 sq mi / 1,964,375 sq km
Population: 114,793,000
Population Density: 151 people per sq mi / 58 people per sq km
Capital: Mexico City

PHYSICAL:

Highest Point: Pico de Orizaba 18,855 ft / 5,747 m
Lowest Point: Laguna Salada 33 ft / 10 m below sea level
Mountain Ranges:
  Sierra Madre: •located in southeastern Mexico •extend into Guatemala
  Sierra Madre del Sur: •located in southern Mexico
  Sierra Madre Occidental: •cover much of western Mexico
  Sierra Madre Oriental: •cover much of eastern Mexico
Peaks (minimum elevation of 8,000 ft / 2,400 m):
  Pico de Orizaba: •located in the Sierra Madre Oriental •highest point in Mexico
  Volcan Popocatepetl: •located in the Sierra Madre Oriental •active volcano
Deserts:
  Chihuahuan: •located in northern Mexico •east of the Sierra Madre Occidental
    •extends into the United States
  Sonoran: •located in northwestern Mexico •west of the Sierra Madre Occidental
    •extends into the United States

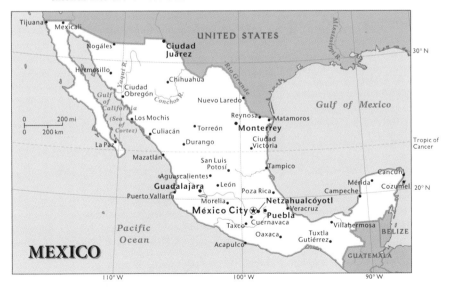

Oceans:
   Pacific: •forms most of the western coast of Mexico •forms part of the southern coast of Mexico
Seas:
   Caribbean: •forms most of the southeastern coast of Mexico •feeds into the Atlantic Ocean
Gulfs:
   Gulf of California: •forms part of the northwestern coast of Mexico •feeds into the Pacific Ocean
   Gulf of Mexico: •forms most of the eastern coast of Mexico •feeds into the Caribbean Sea and Atlantic Ocean
   Gulf of Tehuantepec: •forms part of the southern coast of Mexico •feeds into the Pacific Ocean
Lagoons:
   Laguna Madre: •located in northeastern Mexico •feeds into the Gulf of Mexico
Channels:
   Yucatan: •forms part of the eastern coast of Mexico •separates the Yucatan Peninsula from Cuba •connects the Gulf of Mexico and Caribbean Sea
Lakes:
   Lake Chapala: •located in southwestern Mexico •on the edge of Sierra Madre Occidental
Rivers:
   Balsas: •has its mouth in the Pacific Ocean •has its source in the Sierra Madre Oriental
   Rio Grande: •has its mouth in the Gulf of Mexico •forms much of the Mexico-United States border •extends into the United States
Peninsulas:
   Baja California: •bordered by the Pacific Ocean and Gulf of California
   Yucatan: •bordered by the Caribbean Sea, Gulf of Mexico, and Yucatan Channel

POLITICAL:
Independence: September 16, 1810 (from Spain)
Bordering Countries: United States, Guatemala, Belize (3)
Administrative Divisions: Aguascalientes, Baja California, Baja California Sur, Campeche, Chiapas, Chihuahua, Coahuila, Colima, Durango, Guanajuato, Guerrero, Hidalgo, Jalisco, Mexico, Michoacan de Ocampo, Morelos, Nayarit, Nuevo Leon, Oaxaca, Puebla, Queretaro, Quintana Roo, San Luis Potosi, Sinaloa, Sonora, Tabasco, Tamaulipas, Tlaxcala, Veracruz, Yucatan, Zacatecas (31 states) •Distrito Federal (1 federal district)
Ethnic/Racial Groups: mestizo, Amerindian, white
Religion: Christianity (Roman Catholic, Protestant)
Languages: Spanish, Mayan, Nahuatl, other indigenous languages
Currency: Mexican peso
Current President: Felipe de Jesus Calderon Hinjosa
Cities (capital, largest, or with at least a million people):
   Mexico City: •located in southern Mexico •city in the Distrito Federal and the

Mexico state •capital of Mexico •most populated city in Mexico (19,319,000 people) •on the edge of the Sierra Madre Oriental

Guadalajara: •located in central Mexico •city in the Jalisco state •on the edge of the Sierra Madre Occidental

Puebla: •located in southern Mexico •city in the Puebla state •on the edge of the Sierra Madre Oriental

Ciudad Juarez: •located in northern Mexico •city in the Chihuahua state •chief port on the Rio Grande

Monterrey: •located in central Mexico •city in the Nuevo Leon state •on the edge of the Sierra Madre Oriental

Tijuana: •located in northwestern Mexico •city in the Baja California state •a major port on the Pacific Ocean

Leon: •located in central Mexico •city in the Guanajuato state •on the edge of the Sierra Madre Occidental

ENVIRONMENTAL/ECONOMIC:
Climate: tropical to desert
Natural Resources: petroleum, silver, copper, gold, lead
Agricultural Products: corn, wheat, soybeans, rice, beef
Major Exports: manufactured goods, oil and oil products, silver, fruits, vegetables
Natural Hazards: tsunamis, volcanoes, earthquakes, hurricanes

MICRONESIA
Country Name: Federated States of Micronesia
Continent: Australia/Oceania

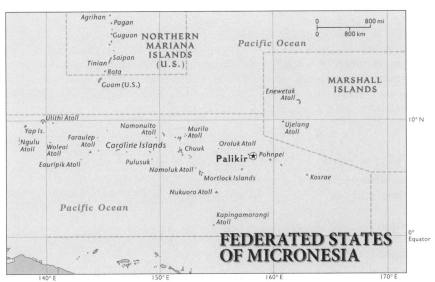

Area: 271 sq mi / 702 sq km
Population: 102,000
Population Density: 376 people per sq mi / 145 people per sq km
Capital: Palikir

PHYSICAL:
Highest Point: Dolohmwar 2,595 ft / 791 m
Lowest Point: Pacific Ocean 0 ft / 0 m (sea level)
Oceans:
  Pacific: •surrounds Micronesia
Islands:
  Caroline: •main island group in Micronesia •surrounded by the Pacific Ocean

POLITICAL:
Independence: November 3, 1986 (from United States administration)
Former Names: Pnape, Truk, and Yap Districts (part of the Trust Territory of the Pacific Islands)
Administrative Divisions: Chuuk, Kosrae, Pohnpei, Yap (4 states)
Ethnic/Racial Groups: Micronesian, Polynesian
Religion: Christianity (Roman Catholic, Protestant)
Languages: English, Trukese, Pohnpeian, Yapese, Kosrean, Ulithian
Currency: U.S. dollar
Current President: Emanuel Mori
Cities (capital, largest, or with at least a million people):
  Palikir: •located in eastern Micronesia •city in the Pohnpei state •capital of Micronesia •most populated city in Micronesia (7,000 people) •chief port on the Pacific Ocean

ENVIRONMENTAL/ECONOMIC:
Climate: tropical wet
Natural Resources: forests, marine products, deep-seabed minerals
Agricultural Products: black pepper, tropical fruits and vegetables, coconuts, cassava (tapioca), pigs
Major Exports: fish, garments, bananas, black pepper
Natural Hazards: typhoons

MOLDOVA
Country Name: Republic of Moldova
Continent: Europe
Area: 13,050 sq mi / 33,800 sq km
Population: 4,109,000
Population Density: 315 people per sq mi / 121 people per sq km
Capital: Chisinau

MOLDOVA

PHYSICAL:
Highest Point: Dealul Balanesti 1,411 ft / 430 m
Lowest Point: Nistru (Dniester) River 7 ft / 2 m
Grasslands/Prairies:
  Black Sea Lowland: •covers much of southern Moldova •extends into Ukraine and Romania
Rivers:
  Nistru (Dniester): •forms part of the Moldova-Ukraine border •extends into Ukraine
  Prut: •forms all of the Moldova-Romania border •extends into Ukraine •tributary of the Danube

POLITICAL:
Independence: August 27, 1991 (from the Soviet Union)
Former Names: Soviet Socialist Republic of Moldova, Moldavia
Bordering Countries: Ukraine, Romania (2)
Regions:
  Bessarabia: •region that covers most of the country •includes part of the Black Sea Lowland •extends into Ukraine
Administrative Divisions: Anenii Noi, Basarabeasca, Briceni, Cahul, Cantemir, Calarasi, Causeni, Cimislia, Criuleni, Donduseni, Drochia, Dubasari, Edinet, Falesti, Floresti, Glodeni, Hincesti, Ialoveni, Leova, Nisporeni, Ocnita, Orhei, Rezina, Riscani, Singerei, Soldanesti, Soroca, Stefan-Voda, Straseni, Taraclia, Telenesti, Ungheni (32 districts, or raions) •Balti, Bender, Chisinau (3 municipalities) •Gagauzia (1 autonomous territorial unit) •Stinga Nistrului, also known as Transdniestria (1 territorial unit)
Ethnic/Racial Groups: Moldovan/Romanian, Ukrainian, Russian

Religion: Christianity (predominantly Eastern Orthodox)
Languages: Moldovan, Russian, Gagauz
Currency: Moldovan leu
Current President: Marian Lupu
Cities (capital, largest, or with at least a million people):
  Chisinau: •located in central Moldova •city in the Chisinau municipality •capital of Moldova •most populated city in Moldova (650,000 people)

ENVIRONMENTAL/ECONOMIC:
Climate: moderate winters/warm summers
Natural Resources: lignite, phosphorites, gypsum, arable land, limestone
Agricultural Products: vegetables, fruits, wine, grain, beef
Major Exports: foodstuffs, textiles, machinery
Natural Hazards: landslides

MONACO
Country Name: Principality of Monaco
Continent: Europe
Area: 1 sq mi / 2 sq km
Population: 36,000
Population Density: 36,000 people per sq mi / 18,000 people per sq km (most dense in Europe and the world)
Capital: Monaco

PHYSICAL:
Highest Point: Mont Agel 459 ft / 140 m
Lowest Point: Mediterranean Sea 0 ft / 0 m (sea level)
Seas:
  Mediterranean: •forms the coast of Monaco •feeds into the Atlantic Ocean

POLITICAL:
Independence: 1419 (beginning of rule by the House of Grimaldi)
Bordering Country: France (1)
Ethnic/Racial Groups: French, Monegasque, Italian
Religion: Christianity (primarily Roman Catholic)
Languages: French, English, Italian, Monegasque
Currency: euro
Current Prince: Albert II
Cities (capital, largest, or with at least a million people):
  Monaco: •makes up the whole country •capital of Monaco •most populated city in Monaco (33,000 people) •chief port on the Mediterranean Sea

ENVIRONMENTAL/ECONOMIC:
Climate: Mediterranean with mild and wet winters/hot and dry summers

MONGOLIA
Country Name: Mongolia
Continent: Asia
Area: 603,909 sq mi / 1,564,116 sq km
Population: 2,814,000
Population Density: 5 people per sq mi / 2 people per sq km (least dense in Asia and the world)
Capital: Ulaanbaatar

PHYSICAL:
Highest Point: Nayramadlin Orgil 14,350 ft / 4,374 m
Lowest Point: Hoh Nuur 1,837 ft / 560 m
Mountain Ranges:
  Altay: •cover an area from northwestern to southern Mongolia •extend into China and Russia
  Eastern Sayans: •located in northern Mongolia •extend into Russia
  Hangayn: •located in central Mongolia
Peaks (minimum elevation of 8,000 ft / 2,400 m):
  Nayramadlin Orgil: •located in the Altay Mountains •on the Mongolia-Russia border •highest point in Mongolia
Plateau Regions:
  Mongolian Plateau: •covers much of eastern Mongolia •extends into China
Deserts:
  Gobi: •covers much of southern Mongolia •extends into China

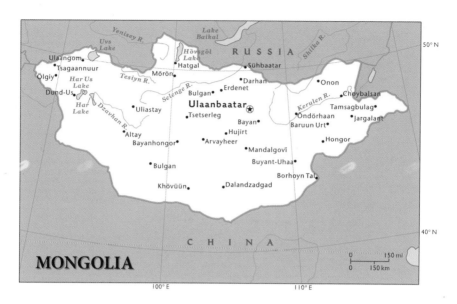

POLITICAL:
Independence: July 11, 1921 (from China)
Former Name: Outer Mongolia
Bordering Countries: China, Russia (2)
Administrative Divisions: Arhangay, Bayanhongor, Bayan-Olgiy, Bulgan, Darhan Uul, Dornod, Dornogovi, Dundgovi, Dzavhan, Govi-Altay, Govisumber, Hentiy, Hovd, Hovsgol, Omnogovi, Orhon, Ovorhangay, Selenge, Suhbaatar, Tov, Uvs (21 provinces) •Ulaanbataar (1 municipality)
Ethnic/Racial Groups: Mongol, Turkic
Religions: Tibetan Buddhism, Lamaism
Languages: Khalkha Mongol, Turkic, Russian
Currency: togrog (tugrik)
Current President: Tsakhia Elbegdorj
Cities (capital, largest, or with at least a million people):
  Ulaanbaatar: •located in northern Mongolia •city in the Ulaanbataar municipality •capital of Mongolia •most populated city in Mongolia (949,000 people)

ENVIRONMENTAL/ECONOMIC:
Climate: desert; continental
Natural Resources: oil, coal, copper, molybdenum, tungsten
Agricultural Products: wheat, barley, potatoes, forage crops, sheep
Major Exports: copper, gold, livestock, animal products, cashmere, wool, hides
Natural Hazards: dust storms, grassland and forest fires, droughts, zud (harsh winter conditions)

MONTENEGRO

Country Name: Montenegro
Continent: Europe
Area: 5,333 sq mi /13,812 sq km
Population: 637,000
Population Density: 119 people per sq mi / 46 people per sq km
Capital: Podgorica

PHYSICAL:
Highest Point: Bobotov Kuk 8,274 ft / 2,522 m
Lowest Point: Adriatic Sea 0 ft / 0 m (sea level)
Mountain Ranges:
   Balkan: •in southeastern Montenegro   •extend into Bulgaria and Serbia
   Dinaric Alps: •extend into Bosnia and Herzegovina
Peaks (minimum elevation of 8,000 ft/2,400 m):
   Bobotov Kuk: •highest point in Montenegro
Grasslands/Prairies:
   Great Alfold: •covers most of northern Serbia, extend into Croatia, Hungary, and Romania
Seas:
   Adriatic Sea: •forms the southwestern coast   •feeds into Ionian Sea
Lakes:
   Lake Scutari: •located in southwestern part of country on the Montenegro-Albanian border

Valleys:
   Zeta River Valley, Tara River Canyon

POLITICAL:
Independence: June 3, 2006 (from the federated union of Serbia and Montenegro; was part of Yugoslavia)
Former Name: Serbia and Montenegro, previously Yugoslavia
Bordering Countries: Albania, Bosnia and Herzegovina, Serbia, Kosovo
Administrative Divisions: Andrijevica, Bar, Berane, Bijelo Polje, Budva, Cetinje, Danilovgrad, Herceg Novi, Kolasin, Kotor, Mojkovac, Niksic, Plav, Pljevlja, Pluzine, Podgorica, Rozaje, Savnik, Tivat, Ulcinj, Zabljak (21 municipalities)
Ethnic/Racial Groups: Montenegrin, Serb, Bosniak, Albanian, Other (Muslims, Croats, Roma)
Religions: Orthodox, Muslim, Roman Catholic
Languages: Montenegrin, Serbian, Bosnian, Albanian
Currency: euro
Current President: Filip Vujanovic
Cities (capital, largest, or with at least a million people):
   Podgorica: •capital of Montenegro   •largest city in Montenegro (144,000 people)

ENVIRONMENTAL/ECONOMIC:
Climate: Mediterranean climate, hot dry summers and autumns, cold winters with substantial snowfall in the inland areas
Natural Resources: bauxite, hydroelectricity
Agricultural: tobacco, potatoes, citrus fruits, olives, sheepherding
Natural Hazards: earthquakes

MOROCCO
Country Name: Kingdom of Morocco
Continent: Africa
Area: 274,461 sq mi / 710,850 sq km
Population: 32,273,000
Population Density: 118 people per sq mi / 45 people per sq km
Capital: Rabat

PHYSICAL:
Highest Point: Jebel Toubkal 13,665 ft / 4,165 m
Lowest Point: Sebkha Tah 180 ft / 55 m below sea level
Mountain Ranges:
   Atlas: •cover an area from southwestern to northeastern Morocco   •extend into Algeria
Peaks (minimum elevation of 8,000 ft / 2,400 m):
   Jebel Toubkal: •located in the Atlas Mountains   •highest point in Morocco

MOROCCO

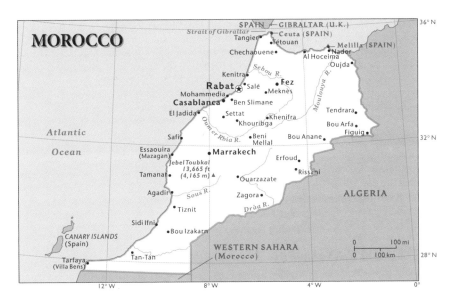

Plateau Regions:
  Hauts Plateaux: •located in northeastern Morocco •extend into Algeria •high basin surrounded by the Atlas Mountains
Oceans:
  Atlantic: •forms the western coast of Morocco
Seas:
  Alboran (sometimes labeled as part of the Mediterranean): •forms the northeastern coast of Morocco •feeds into the Atlantic Ocean
Straits:
  Strait of Gibraltar: •forms part of the northern coast of Morocco •separates Morocco from Spain •connects the Alboran Sea and Atlantic Ocean

POLITICAL:

Independence: March 2, 1956 (from France)
Bordering Countries: Algeria, Western Sahara (claimed by Morocco) (2)
Administrative Divisions: Grand Casablanca, Chaouia-Ouardigha, Doukkala-Abda, Fes-Boulemane, Gharb-Chrarda-Beni Hssen, Guelmim-Es Smara, Laayoune-Boujdour-Sakia El Hamra, Marrakech-Tensift-Al-Haouz, Meknes-Tafilalet, Oriental, Rabat-Sale-Zemmour-Zaer, Souss-Massa-Draa, Tadla-Azilal, Tanger-Tetouan, Taza-Al Hoceima-Taounate (15 regions)
Ethnic/Racial Groups: predominantly Arab-Berber
Religion: predominantly Islam
Languages: Arabic, Berber dialects, French
Currency: Moroccan dirham
Current King: Mohammed VI

Cities (capital, largest, or with at least a million people):
Casablanca: •located in northwestern Morocco •city in the Casablanca wilaya •most populated city in Morocco (3,284,000 people) •a major port on the Atlantic Ocean
Rabat: •located in northwestern Morocco •city in the Rabat-Sale wilaya •capital of Morocco (1,770,000 people)
Fez: •located in northern Morocco •city in the Fez province •on the edge of the Atlas Mountains

ENVIRONMENTAL/ECONOMIC:
Climate: Mediterranean; more extreme in the mountains
Natural Resources: phosphates, iron ore, manganese, lead, zinc
Agricultural Products: barley, wheat, citrus, wine, livestock
Major Exports: clothing, fish, inorganic chemicals, transistors, crude materials
Natural Hazards: earthquakes, droughts

MOZAMBIQUE
Country Name: Republic of Mozambique
Continent: Africa
Area: 308,642 sq mi / 799,380 sq km
Population: 23,050,000
Population Density: 75 people per sq mi / 29 people per sq km
Capital: Maputo

PHYSICAL:
Highest Point: Monte Binga 7,992 ft / 2,436 m

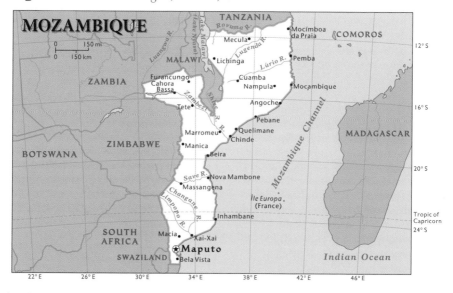

Lowest Point: Indian Ocean 0 ft / 0 m (sea level)
Valleys:
    Great Rift: •located in central Mozambique •extends into Malawi and Tanzania
Oceans:
    Indian: •forms part of the northeastern coast of Mozambique •forms much of the southern coast of Mozambique
Channels:
    Mozambique: •forms most of the eastern coast of Mozambique •separates Mozambique and Madagascar
Lakes:
    Lake Cahora Bassa: •located in northwestern Mozambique •fed by the Zambezi River •formed by the Cahora Bassa Dam
    Lake Malawi: •located in northern Mozambique •located in the Great Rift Valley •on the border with Malawi •extends into Tanzania
Rivers:
    Chire (Shire): •tributary of the Zambezi River •extends into Malawi
    Limpopo: •has its mouth in the Indian Ocean •extends into South Africa and Zimbabwe
    Luangwa: •tributary of the Zambezi River •forms part of the Mozambique-Zambia border •extends into Zambia
    Save: •has its mouth in the Mozambique Channel •extends into Zimbabwe
    Zambeze (Zambezi): •has its mouth in the Mozambique Channel •extends into Zambia

POLITICAL:

Independence: June 25, 1975 (from Portugal)
Former Name: Portuguese East Africa
Bordering Countries: Malawi, Zimbabwe, Tanzania, South Africa, Zambia, Swaziland (6)
Administrative Divisions: Cabo Delgado, Gaza, Inhambane, Manica, Maputo, Nampula, Niassa, Sofala, Tete, Zambezia (10 provinces) •Maputo (1 city)
Ethnic/Racial Groups: Shangaan, Chockwe, Manyika, Sena, Makua
Religions: indigenous beliefs, Christianity, Islam
Languages: Portuguese, indigenous dialects
Currency: metical
Current President: Armando Guebuza
Cities (capital, largest, or with at least a million people):
    Maputo: •located in southern Mozambique •city in the Maputo province •capital of Mozambique •most populated city in Mozambique (1,589,000 people) •chief port on the Indian Ocean

ENVIRONMENTAL/ECONOMIC:

Climate: tropical to subtropical
Natural Resources: coal, titanium, natural gas, hydropower, tantalum
Agricultural Products: cotton, cashew nuts, sugarcane, tea, beef
Major Exports: aluminum, prawns, cashews, cotton, sugar
Natural Hazards: droughts, cyclones, floods

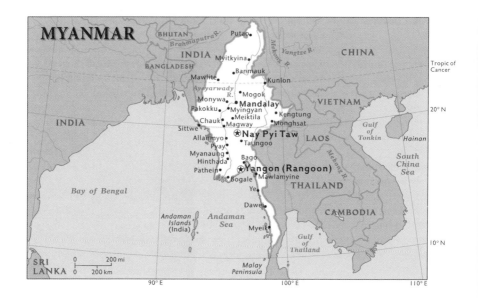

MYANMAR

Country Name: Union of Myanmar
Continent: Asia
Area: 261,218 sq mi / 676,552 sq km
Population: 54,000,000
Population Density: 207 people per sq mi / 80 people per sq km
Capital: Nay Pyi Taw

PHYSICAL:
Highest Point: Hkakabo Razi 19,295 ft / 5,881 m
Lowest Point: Andaman Sea 0 ft / 0 m (sea level)
Mountain Ranges:
  Hengduan Shan: •located in northern Myanmar •extend into China
Peaks (minimum elevation of 8,000 ft / 2,400 m):
  Hkakabo Razi: •located in the Hengduan Shan •on the Myanmar-China border •highest point in Myanmar
Plateau Regions:
  Shan Plateau: •located in eastern Myanmar
Seas:
  Andaman: •forms most of the southern coast of Myanmar •feeds into the Bay of Bengal and Indian Ocean
Bays:
  Bay of Bengal: •forms most of the western coast of Myanmar •feeds into the Indian Ocean

Rivers:
  Ayeyarwady: •has its mouth in the Andaman Sea
  Mekong: •forms all of the Myanmar-Laos border •extends into China and Laos
  Salween (Thanlwin): •has its mouth in the Andaman Sea •forms part of the Myanmar-Thailand border •extends into China
Deltas:
  Mouths of the Ayeyarwady: •mouth of the Ayeyarwady River •feeds into the Andaman Sea

POLITICAL:
Independence: January 4, 1948 (from the United Kingdom)
Former Names: Burma, Socialist Republic of the Union of Burma
Bordering Countries: China, Thailand, India, Laos, Bangladesh (5)
Administrative Divisions: Chin, Kachin, Kayin, Kayah, Mon, Rakhine, Shan (7 states) •Ayeyarwady, Bago, Magway, Mandalay, Sagaing, Tanintharyi, Yangon (7 districts)
Ethnic/Racial Groups: Burman, Shan, Karen, Rahkine, Chinese
Religions: Buddhism, Christianity, Islam
Languages: Burmese, minor languages
Currency: kyat
Current Chairman: Than Shwe
Cities (capital, largest, or with at least a million people):
  Yangon: •located in central Myanmar •city in the Yangon district • most populated city in Myanmar (4,350,000 people) •chief port on the Andaman Sea
  Mandalay: •located in central Myanmar •city in the Mandalay district •chief port on the Ayeyarwady River
  Nay Pyi Taw: •located in south-central Myanmar •capital of Myanmar (1,024,000 people)

ENVIRONMENTAL/ECONOMIC:
Climate: tropical monsoon with hot and rainy summers/mild winters
Natural Resources: petroleum, timber, tin, antimony, zinc
Agricultural Products: rice, pulses, beans, sesame, hardwood (teak)
Major Exports: natural gas, wood products, pulses, beans, fish
Natural Hazards: earthquakes, cyclones, floods, landslides, droughts

NAMIBIA

Country Name: Republic of Namibia
Continent: Africa
Area: 318,261 sq mi / 824,292 sq km
Population: 2,324,000
Population Density: 7 people per sq mi / 3 people per sq km (least dense in Africa)
Capital: Windhoek

PHYSICAL:

Highest Point: Konigstein 8,550 ft / 2,606 m
Lowest Point: Atlantic Ocean 0 ft / 0 m (sea level)
Mountain Ranges:
    Kaokoveld: •located in northwestern Namibia
Peaks (minimum elevation of 8,000 ft / 2,400 m):
    Brandberg: •located in the Kaokoveld •highest point in Namibia
Deserts:
    Kalahari: •covers much of eastern Namibia •extends into Botswana and South Africa
    Namib: •located along the western coast of Namibia

Oceans:
　Atlantic: •forms the western coast of Namibia
Rivers:
　Okavango: •forms part of the Namibia-Angola border •extends into Angola and Botswana
　Orange: •has its mouth in the Atlantic Ocean •forms part of the Namibia-South Africa border •extends into South Africa
Regions:
　Kaukau Veld: •lowland region in northeastern Namibia •extends into Botswana •north of the Kalahari Desert
　Skeleton Coast: •region in western Namibia •includes part of the Atlantic coast •west of the Namib Desert

POLITICAL:
Independence: March 21, 1990 (from South African administration)
Former Names: German Southwest Africa, South-West Africa
Bordering Countries: Angola, Botswana, South Africa, Zambia (4)
Regions:
　Caprivi Strip: •region in northeastern Namibia •panhandle bordered on the north by Angola and Zambia and on the south by Botwana •extends as far east as Zimbabwe
　Damaraland: •region that covers much of central Namibia •includes part of the Kaokoveld and the Namib Desert
　Great Namaland: •region that covers much of southern Namibia •includes part of the Namib Desert
　Ovamboland: •region in northern Namibia •includes part of the Kaokoveld
Administrative Divisions: Caprivi, Erongo, Hardap, Karas, Khomas, Kunene, Ohangwena, Okavango, Omaheke, Omusati, Oshana, Oshikoto, Otjozondjupa (13 regions)
Ethnic/Racial Groups: Ovambo, Kavangos, Herero, Damara, white
Religions: Christianity, indigenous beliefs
Languages: English, Afrikaans, German, indigenous languages
Currency: Namibian dollar, South African rand
Current President: Hifikepunye Pohamba
Cities (capital, largest, or with at least a million people):
　Windhoek: •located in central Namibia •city in the Khomas region •capital of Namibia •most populated city in Namibia (342,000 people)

ENVIRONMENTAL/ECONOMIC:
Climate: desert, hot and dry
Natural Resources: diamonds, copper, uranium, gold, lead
Agricultural Products: millet, sorghum, peanuts, livestock, fish
Major Exports: diamonds, copper, gold, zinc, lead
Natural Hazards: droughts

NAURU

Country Name: Republic of Nauru
Continent: Australia/Oceania
Area: 8 sq mi / 21 sq km (smallest in Australia/Oceania)
Population: 10,000
Population Density: 1,250 people per sq mi / 476 people per sq km (most dense in Australia/Oceania)
Capital: No official capital. Government offices are in Yaren.

PHYSICAL:

Highest Point: unnamed location 200 ft / 61 m
Lowest Point: Pacific Ocean 0 ft / 0 m (sea level)
Oceans:
  Pacific: •surrounds Nauru

POLITICAL:

Independence: January 31, 1968 (from Australia/New Zealand/United Kingdom administered UN trusteeship)
Former Name: Pleasant Island
Administrative Divisions: Aiwo, Anabar, Anetan, Anibare, Baiti, Boe, Buada, Denigomodu, Ewa, Ijuw, Meneng, Nibok, Uaboe, Yaren (14 districts)
Ethnic/Racial Groups: Nauruan, other Pacific Islander, Chinese, European
Religion: Christianity (Protestant, Roman Catholic)
Languages: Nauruan, English
Currency: Australian dollar

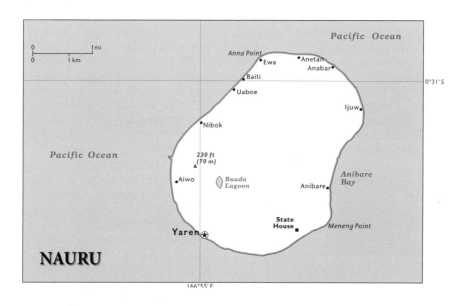

Current President: Marcus Stephen
Cities (capital, largest, or with at least a million people):
Yaren: •located in southern Nauru •city in the Yaren district •unofficial capital of Nauru •most populated city in Nauru (9,000 people) •chief port on the Pacific Ocean

ENVIRONMENTAL/ECONOMIC:
Climate: tropical; monsoonal with rainy/dry seasons
Natural Resources: phosphates, fish
Agricultural Products: coconuts
Major Exports: phosphates
Natural Hazards: droughts

NEPAL
Country Name: Federal Democratic Republic of Nepal
Continent: Asia
Area: 56,827 sq mi / 147,181 sq km
Population: 30,486,000
Population Density: 536 people per sq mi / 207 people per sq km
Capital: Kathmandu

PHYSICAL:
Highest Point: Mount Everest 29,035 ft / 8,850 m
Lowest Point: Kanchan Kalan 230 ft / 70 m

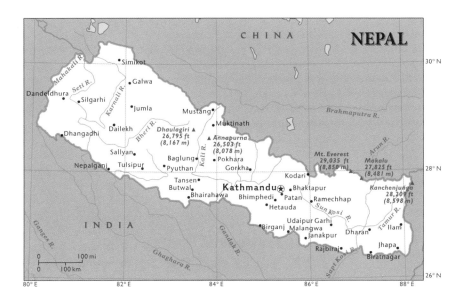

Mountain Ranges:
   Himalaya: •cover much of northern Nepal •extend into China and India
Peaks (minimum elevation of 8,000 ft / 2,400 m):
   Annapurna: •located in the Himalaya
   Cho Oyu: •located in the Himalaya •on the Nepal-China border
   Dhaulagiri: •located in the Himalaya
   Everest: •located in the Himalaya •on the Nepal-China border •highest point in the world
   Kanchenjunga: •located in the Himalaya •on the Nepal-India border
   Lhotse: •located in the Himalaya
   Makalu •located in the Himalaya •on the Nepal-China border
   Manaslu: •located in the Himalaya

POLITICAL:
Independence: 1768 (unified by Prithvi Narayan Shah)
Bordering Countries: India, China (2)
Administrative Divisions: Bagmati, Bheri, Dhawalagiri, Gandaki, Janakpur, Karnali, Kosi, Lumbini, Mahakali, Mechi, Narayani, Rapti, Sagarmatha, Seti (14 zones)
Ethnic/Racial Groups: Brahman, Chetri, Newar, Gurung, Magar, Tamang, Rai, Limbu, Sherpa, Tharu
Religions: Hinduism, Buddhism, Islam
Languages: Nepali, English, many other languages and dialects
Currency: Nepalese rupee
Current Prime Minister: Baburam Bhattarai
Cities (capital, largest, or with at least a million people):
   Kathmandu: •located in central Nepal •city in the Bagmati zone •capital of Nepal •most populated city in Nepal (990,000 people) •on the edge of the Himalaya

ENVIRONMENTAL/ECONOMIC:
Climate: mountainous with cool summers and harsh winters in the north; subtropical with mild winters in the south
Natural Resources: quartz, water, timber, hydropower, scenic beauty
Agricultural Products: rice, corn, wheat, sugarcane, milk
Major Exports: carpets, clothing, leather goods, jute goods, grain
Natural Hazards: severe thunderstorms, floods, landslides, drought, long monsoon seasons

NETHERLANDS
Country Name: Kingdom of the Netherlands
Continent: Europe
Area: 16,034 sq mi / 41,528 sq km
Population: 16,694,000
Population Density: 1,041 people per sq mi / 402 people per sq km
Capital: Amsterdam

PHYSICAL:
Highest Point: Vaalserberg 1,053 ft / 321 m
Lowest Point: Zuidplaspolder 23 ft / 7 m below sea level
Seas:
  North: •forms much of the western and northern coasts of the Netherlands •feeds into the Atlantic Ocean
  Wadden: •forms part of the northern coast of the Netherlands •feeds into the North Sea
Lakes:
  IJsselmeer: •located in northwestern Netherlands
Rivers:
  Maas: •has its mouth in the North Sea •forms part of the Netherlands-Belgium border •extends into Belgium
  Rhine: •has its mouth in the North Sea •extends into Germany
  Schelde: •has its mouth in the Westerschelde estuary •extends into Belgium
Deltas:
  Westerschelde estuary: •mouth of the Schelde River •feeds into the North Sea
Islands:
  Flevoland: •island in IJsselmeer Lake
  West Frisian: •island chain off the northern coast of the Netherlands •bordered by the North and Wadden Seas

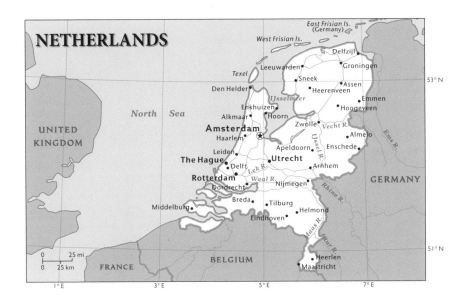

POLITICAL:
Independence: 1579 (from Spain)
Bordering Countries: Germany, Belgium (2)
Administrative Divisions: Drenthe, Flevoland, Friesland, Gelderland, Groningen, Limburg, North Brabant, North Holland, Overijssel, South Holland, Utrecht, Zeeland, (12 provinces)
Ethnic/Racial Groups: predominantly Dutch
Religions: Christianity (Roman Catholic, Protestant), Islam
Languages: Dutch, Frisian
Currency: euro
Current Queen: Beatrix
Cities (capital, largest, or with at least a million people):
  Amsterdam: •located in western Netherlands •city in the Noord-Holland province •capital of the Netherlands •most populated city in the Netherlands (1,044,000 people) •chief port on IJsselmeer Lake

ENVIRONMENTAL/ECONOMIC:
Climate: temperate marine with cool summers/mild winters
Natural Resources: natural gas, petroleum, peat, limestone, salt
Agricultural Products: grains, potatoes, sugar beets, fruits, livestock
Major Exports: machinery and equipment, chemicals, fuels, foodstuffs
Natural Hazards: floods

NEW ZEALAND
Country Name: New Zealand
Continent: Australia/Oceania
Area: 104,454 sq mi / 270,534 sq km
Population: 4,417,000
Population Density: 42 people per sq mi / 16 people per sq km
Capital: Wellington

PHYSICAL:
Highest Point: Mount Cook 12,316 ft / 3,754 m
Lowest Point: Pacific Ocean 0 ft / 0 m (sea level)
Mountain Ranges:
  Southern Alps: •cover much of New Zealand •located on South Island
Peaks (minimum elevation of 8,000 ft / 2,400 m):
  Mount Cook: •located in the Southern Alps •highest point in New Zealand
  Mount Ruapehu: •located in the Southern Alps
Oceans:
  Pacific: •forms most of the eastern coast of New Zealand
Seas:
  Tasman: •forms most of the western coast of New Zealand •feeds into the Pacific Ocean

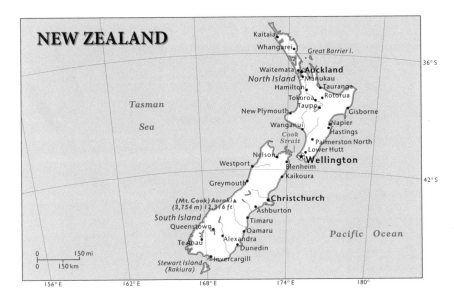

Bays:
  Bay of Plenty: •forms part of the northern coast of North Island •feeds into the Pacific Ocean
  Canterbury Bight: •forms part of the eastern coast of South Island •feeds into the Pacific Ocean
Straits:
  Cook: •forms much of the inner coast of New Zealand •separates North and South Islands •connects the Tasman Sea and Pacific Ocean
Islands:
  North: •second largest island in New Zealand •bordered by the Pacific Ocean, Tasman Sea, Bay of Plenty, and Cook Strait
  South: •largest island in New Zealand •bordered by the Pacific Ocean, Tasman Sea, Canterbury Bight, and Cook Strait

POLITICAL:
Independence: September 26, 1907 (from the United Kingdom)
Administrative Divisions: Auckland, Bay of Plenty, Canterbury, Gisborne, Hawke's Bay, Manawatu-Wanganui, Marlborough, Nelson, Northland, Otago, Southland, Taranaki, Tasman, Waikato, Wellington, West Coast (16 regions) •Chatham Islands (1 territory)
Ethnic/Racial Groups: New Zealand European (Kiwi), Maori (indigenous), other European, Pacific Islander
Religion: Christianity (Protestant, Roman Catholic)
Languages: English, Maori

Currency: New Zealand dollar
Current Prime Minister: John Phillip Key
Cities (capital, largest, or with at least a million people):
　Auckland:　•located on North Island •city in the Auckland region •most populated city in New Zealand (1,404,000 people) •chief port on the Pacific Ocean and Tasman Sea
　Wellington:　•located on North Island •city in the Wellington region •capital city (391,000 people) •chief port on the Cook Strait

ENVIRONMENTAL/ECONOMIC:
Climate: temperate
Natural Resources: natural gas, iron ore, sand, coal, timber
Agricultural Products: wheat, barley, potatoes, pulses, wool
Major Exports: dairy products, meat, wood and wood products, fish, chemicals
Natural Hazards: earthquakes, volcanoes

NICARAGUA
Country Name: Republic of Nicaragua
Continent: North America
Area: 50,193 sq mi / 130,000 sq km
Population: 5,870,000
Population Density: 117 people per sq mi / 45 people per sq km
Capital: Managua

PHYSICAL:
Highest Point: Pico Mogotan 7,999 ft / 2,438 m
Lowest Point: Pacific Ocean 0 ft / 0 m (sea level)
Oceans:
    Pacific: • forms most of the western coast of Nicaragua
Seas:
    Caribbean: • forms the eastern coast of Nicaragua • feeds into the Atlantic Ocean
Gulfs:
    Gulf of Fonseca: • forms part of the northwestern coast of Nicaragua • feeds into the Pacific Ocean
Lakes:
    Lake Managua: • located in western Nicaragua
    Lake Nicaragua: • located in southwestern Nicaragua • largest lake in Central America
Regions:
    Mosquito Coast: • lowland region in eastern Nicaragua • borders the Caribbean coast • extends into Honduras

POLITICAL:
Independence: September 15, 1821 (from Spain)
Bordering Countries: Honduras, Costa Rica (2)
Administrative Divisions: Boaco, Carazo, Chinandega, Chontales, Esteli, Granada, Jinotega, Leon, Madriz, Managua, Masaya, Matagalpa, Nueva Segovia, Rio San Juan, Rivas (15 departments) • Atlantico Norte, Atlantico Sur (2 autonomous regions)
Ethnic/Racial Groups: mestizo, white, black, Amerindian
Religion: Christianity (Roman Catholic, Protestant)
Languages: Spanish, English, indigenous languages
Currency: gold cordoba
Current President: Daniel Ortega Saavedra
Cities (capital, largest, or with at least a million people):
    Managua: • located in western Nicaragua • city in the Managua department • capital of Nicaragua • most populated city in Nicaragua (934,000 people) • chief port on Lake Managua

ENVIRONMENTAL/ECONOMIC:
Climate: tropical in the lowlands; cooler in the highlands
Natural Resources: gold, silver, copper, tungsten, lead
Agricultural Products: coffee, bananas, sugarcane, cotton, beef
Major Exports: coffee, shrimp and lobster, cotton, tobacco, bananas
Natural Hazards: earthquakes, volcanoes, landslides, hurricanes

NIGER

Country Name: Republic of Niger
Continent: Africa
Area: 489,191 sq mi / 1,267,000 sq km
Population: 16,069,000
Population Density: 33 people per sq mi / 13 people per sq km
Capital: Niamey

PHYSICAL:

Highest Point: Monts Bagzane 6,634 ft / 2,022 m
Lowest Point: Niger River 656 ft / 200 m
Mountain Ranges:
  Air: •located in central Niger
Grasslands/Prairies:
  Sahel: •covers much of southern Niger •extends into Mali, Burkina Faso, Benin, Nigeria, and Chad
Plateau Regions:
  Plateau of Djado: •located in northeastern Niger •desert basin in the Sahara
Deserts:
  Sahara: •covers much of northern and central Niger •extends into Mali, Algeria, Libya, and Chad •largest desert in the world
  Talak: •located in western Niger •extends into Mali and Algeria
  Tenere: •covers an area from northern to southeastern Niger •extends into Algeria
Rivers:
  Niger: •forms part of the Niger-Benin border •extends into Mali and Nigeria

POLITICAL:
Independence: August 3, 1960 (from France)
Bordering Countries: Nigeria, Chad, Algeria, Mali, Burkina Faso, Libya, Benin (7)
Administrative Divisions: Agadez, Diffa, Dosso, Maradi, Tahoua, Tillaberi, Zinder (7 departments) •Niamey (1 capital district)
Ethnic/Racial Groups: Hausa, Djerma, Fula, Tuareg, Beri Beri (Kanouri)
Religions: Islam, indigenous beliefs, Christianity
Languages: French, Hausa, Djerma
Currency: Communaute Financiere Africaine franc
Current President: Issoufou Mahamadou
Cities (capital, largest, or with at least a million people):
    Niamey: •located in southwestern Niger •city in the Niamey capital district •capital of Niger •most populated city in Niger (1,004,000 people) •chief port on the Niger River

ENVIRONMENTAL/ECONOMIC:
Climate: desert, hot and dry; tropical in the far south
Natural Resources: uranium, coal, iron ore, tin, phosphates
Agricultural Products: cowpeas, cotton, peanuts, millet, cattle
Major Exports: uranium ore, livestock, cowpeas, onions
Natural Hazards: droughts

NIGERIA
Country Name: Federal Republic of Nigeria
Continent: Africa
Area: 356,669 sq mi / 923,768 sq km
Population: 162,265,000 (largest in Africa)
Population Density: 455 people per sq mi / 176 people per sq km
Capital: Abuja

PHYSICAL:
Highest Point: Chappal Wadi 7,936 ft / 2,419 m
Lowest Point: Gulf of Guinea 0 ft / 0 m (sea level)
Mountain Ranges:
    Adamawa: •located in southeastern Nigeria •extends into Cameroon
Grasslands/Prairies:
    Sahel: •covers much of northern Nigeria •extends into Benin, Niger, Chad, and Cameroon
    Sokoto Plains: •located in northwestern Nigeria
Plateau Regions:
    Jos: •located in central Nigeria
    Plateau of Yorubaland: •located in western Nigeria

Gulfs:
 Gulf of Guinea: •forms part of the southern coast of Nigeria •feeds into the Atlantic Ocean
Bays:
 Bight of Benin: •forms the southwestern coast of Nigeria •feeds into the Gulf of Guinea
 Bight of Bonny: •forms the southeastern coast of Nigeria •feeds into the Gulf of Guinea
Lakes:
 Kainji Reservoir: •located in western Nigeria •formed by the Kainji Dam •fed by the Niger River
Rivers:
 Benue: •tributary of the Niger River •extends into Cameroon
 Niger: •has its mouth in the Gulf of Guinea •extends into Niger
Deltas:
 Niger Delta: •mouth of the Niger River •ends in the Gulf of Guinea, the Bight of Benin, and the Bight of Bonny
Dams:
 Kainji: •on the Niger River •forms Kainji Reservoir •in western Nigeria

POLITICAL:
Independence: October 1, 1960 (from the United Kingdom)
Bordering Countries: Cameroon, Niger, Benin, Chad (4)
Administrative Divisions: Abia, Adamawa, Akwa Ibom, Anambra, Bauchi, Bayelsa, Benue, Borno, Cross River, Delta, Ebonyi, Edo, Ekiti, Enugu, Gombe, Imo, Jigawa,

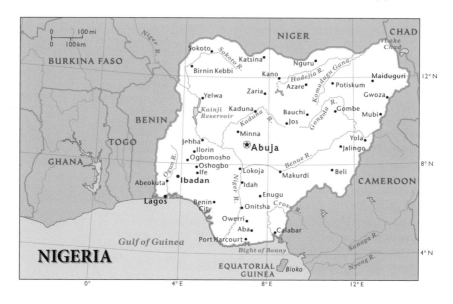

Kaduna, Kano, Katsina, Kebbi, Kogi, Kwara, Lagos, Nassarawa, Niger, Ogun, Ondo, Osun, Oyo, Plateau, Rivers, Sokoto, Taraba, Yobe, Zamfara (36 states) •Abuja Federal Capital Territory (1 territory)

Ethnic/Racial Groups: more than 250 native African groups
Religions: Islam, Christianity, indigenous beliefs
Languages: English, Hausa, Yoruba, Igbo, Fulani
Currency: naira
Current President: Goodluck Jonathan
Cities (capital, largest, or with at least a million people):
  Lagos: •located in southwestern Nigeria •city in the Lagos state •most populated city in Nigeria (9,954,000 people) •chief port on the Bight of Benin
  Ibadan: •located in southwestern Nigeria •city in the Oyo state •on the edge of the Plateau of Yorubaland
  Kano: •located in northern Nigeria •city in the Kano state • located in the Sahel
  Kaduna: •located in central Nigeria •city in the Kaduna state
  Port Harcourt: •located in southern Nigeria •city in the Rivers state
  Benin: •located in southwestern Nigeria •city in the Edo state
  Maiduguri: •located in northeastern Nigeria •city in the Borno state
  Abuja: •located in central Nigeria •city in the Abuja Federal Capital Territory •capital of Nigeria (1,857,000 people)

ENVIRONMENTAL/ECONOMIC:
Climate: equatorial in the south; tropical in the center; arid in the north
Natural Resources: natural gas, petroleum, tin, columbite, iron ore
Agricultural Products: cacao, peanuts, palm oil, corn, cattle
Major Exports: petroleum and petroleum products, cacao, rubber
Natural Hazards: droughts, floods

NORWAY
Country Name: Kingdom of Norway
Continent: Europe
Area: 125,004 sq mi / 323,758 sq km
Population: 4,952,000
Population Density: 40 people per sq mi / 15 people per sq km
Capital: Oslo

PHYSICAL:
Highest Point: Galdhopiggen 8,100 ft / 2,469 m
Lowest Point: Norwegian Sea 0 ft / 0 m (sea level)
Mountain Ranges:
  Kjolen: •cover much of Norway •extend into Sweden and Finland

Peaks (minimum elevation of 8,000 ft / 2,400 m):
  Galdhopiggen: •located in the Kjolen •highest point in Norway
Plateau Regions:
  Maanselka: •located in northeastern Norway •extends into Sweden and Finland
    •large basin west of the Kjolen
Seas:
  Barents: •forms part of the northeastern coast of Norway •feeds into the Arctic Ocean
  North: •forms much of the southwestern coast of Norway •feeds into the
    Atlantic Ocean
  Norwegian: •forms most of the western coast of Norway •feeds into the Atlantic
    and Arctic Oceans
Channels:
  Skagerrak: •forms much of the southern coast of Norway •separates Norway from
    Denmark •connects the Kattegat and North Sea

POLITICAL:
Independence: June 7, 1905 (from Sweden)
Bordering Countries: Sweden, Finland, Russia (3)
Regions:
  Lapland: •region in northern Norway •includes part of the Kjolen •extends into
    Sweden and Finland
Administrative Divisions: Akershus, Aust-Agder, Buskerud, Finnmark, Hedmark,
  Hordaland, More og Romsdal, Nordland, Nord-Trondelag, Oppland, Oslo, Ostfold,
  Rogaland, Sogn og Fjordane, Sor-Trondelag, Telemark, Troms, Vest-Agder, Vestfold
  (19 counties)

External Territories: Bouvet Island, Jan Mayen, Svalbard
Ethnic/Racial Groups: Norwegian, Sami
Religion: Christianity (Evangelical Lutheran)
Languages: Norwegian
Currency: Norwegian krone
Current President: King Harald V
Cities (capital, largest, or with at least a million people):
    Oslo: •located in southeastern Norway •city in the Oslo province •capital of Norway •most populated city in Norway (875,000 people) •southeast of the Kjolen •chief port on the Skagerrak

ENVIRONMENTAL/ECONOMIC:
Climate: temperate along the coast, moderated by North Atlantic Current; cooler and rainier in the interior
Natural Resources: petroleum, natural gas, iron ore, copper, lead
Agricultural Products: barley, wheat, potatoes, pork, fish
Major Exports: petroleum and petroleum products, machinery and equipment, metals, chemicals, ships
Natural Hazards: rockslides, avalanches

OMAN
Country Name: Sultanate of Oman
Continent: Asia
Area: 119,500 sq mi / 309,500 sq km
Population: 2,997,000
Population Density: 25 people per sq mi / 10 people per sq km
Capital: Muscat

PHYSICAL:
Highest Point: Jabal Shams 9,777 ft / 2,980 m
Lowest Point: Arabian Sea 0 ft / 0 m (sea level)
Peaks (minimum elevation of 8,000 ft / 2,400 m):
    Jabal Shams: •located in northern Oman •highest point in Oman
Deserts:
    Rub al Khali: •located in western Oman •extends into Saudi Arabia, Yemen, and United Arab Emirates
Seas:
    Arabian: •forms most of the southern and eastern coasts of Oman •feeds into the Indian Ocean
Gulfs:
    Gulf of Oman: •forms most of the northern coast of Oman •feeds into the Arabian Sea
    Persian: •forms part of the northwestern coast of Oman •feeds into the Gulf of Oman
Straits:
    Strait of Hormuz: •forms part of the northwestern coast of Oman •separates Oman from Iran •connects the Persian Gulf and Gulf of Oman

POLITICAL:
Independence: 1650 (from Portugal)
Former Name: Muscat and Oman
Bordering Countries: Saudi Arabia, United Arab Emirates, Yemen (3)
Administrative Divisions: Ad Dakhiliyah, Al Batinah, Al Wusta, Ash Sharqiyah, Az Zahirah (5 regions) • Al Buraymi, Masqat (Muscat), Musandam, Zufar (4 governorates)
Ethnic/Racial Groups: Arab, Baluchi, South Asian, African
Religions: Islam (Ibadhi, Sunni, Shiite), Hinduism
Languages: Arabic, English, Baluchi, Urdu, Indian dialects
Currency: Omani rial

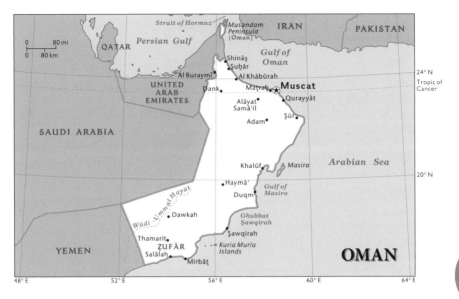

Current Sultan: Qaboos bin Said Al-Said
Cities (capital, largest, or with at least a million people):
  Muscat: •located in northeastern Oman •city in the Masqat region •capital of Oman •most populated city in Oman (634,000 people) •chief port on the Gulf of Oman

ENVIRONMENTAL/ECONOMIC:
Climate: dry desert; hot and humid along the coast, hot and dry in the interior; summer monsoon in the far south
Natural Resources: petroleum, copper, asbestos, marble, limestone
Agricultural Products: dates, limes, bananas, alfalfa, camels
Major Exports: petroleum, re-exports, fish, metals, textiles
Natural Hazards: sandstorms, dust storms, droughts

COUNTRIES

PAKISTAN

Country Name: Islamic Republic of Pakistan
Continent: Asia
Area: 307,374 sq mi / 796,095 sq km
Population: 176,940,000
Population Density: 576 people per sq mi / 222 people per sq km
Capital: Islamabad

PHYSICAL:
Highest Point: K2 28,251 ft / 8,611 m
Lowest Point: Arabian Sea 0 ft / 0 m (sea level)
Mountain Ranges:
- Central Makran: •located in southwestern Pakistan
- Himalaya: •located in northeastern Pakistan •extend into China and India
- Hindu Kush: •cover parts of northern Pakistan •extend into Afghanistan
- Karakoram: •located in northeastern Pakistan •extend into China and India

Peaks (minimum elevation of 8,000 ft / 2,400 m):
- Broad Peak: •located in the Karakoram Range
- Gasherbrum I: •located in the Karakoram Range
- Gasherbrum II: •located in the Karakoram Range

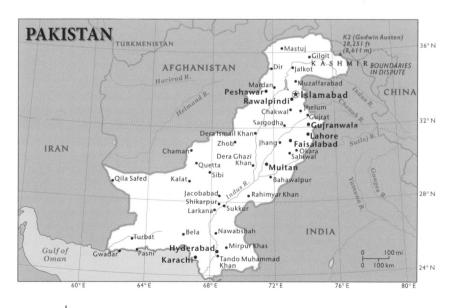

PAGE 258 | THE ULTIMATE FACT BOOK

K2: •located in the Karakoram Range •on the Pakistan-China border •highest point in Pakistan •also called Godwin Austen
Nanga Parbat: •located in the Himalaya
Passes:
Khyber: •located in the Safed Koh Range in the Hindu Kush •crosses from Pakistan into Afghanistan
Grasslands/Prairies:
Punjab Plains: •located in eastern Pakistan •extend into India
Deserts:
Thar: •covers much of eastern Pakistan •extends into India
Seas:
Arabian: •forms most of the southern coast of Pakistan •feeds into the Indian Ocean
Rivers:
Indus: •has its mouth in the Arabian Sea •extends into India and China
Deltas:
Mouths of the Indus: •mouth of the Indus River •feeds into the Arabian Sea

POLITICAL:
Independence: August 14, 1947 (from the United Kingdom)
Former Name: West Pakistan
Bordering Countries: India, Afghanistan, Iran, China (4)
Regions:
Kashmir: •located in northeastern Pakistan •includes parts of the Himalaya, the Hindu Kush, and the Karakoram Range •extends into India
Baluchistan: •located in southwestern Pakistan •includes most of the Central Makran Range •extends into Iran
Administrative Divisions: Balochistan, Khyber Pakhtunkhwa, Punjab, Sindh (4 provinces) •Federally Administered Tribal Areas (1 territory) •Islamabad Capital Territory (1 capital territory) • Azad Kashmir, Gilgit-Baltistan (tribal areas)
Ethnic/Racial Groups: Punjabi, Sindhi, Pashtun, Baloch
Religions: Islam, Christianity, Hinduism
Languages: Punjabi, Sindhi, Siraiki, Pashtu, Urdu, English
Currency: Pakistani rupee
Current President: Asif Ali Zardari
Cities (capital, largest, or with at least a million people):
Karachi: •located in southern Pakistan •city in the Sindh province •most populated city in Pakistan (13,125,000 people) •chief port on the Arabian Sea
Lahore: •located in eastern Pakistan •city in the Punjab province •in the center of the Punjab Plains
Faisalabad: •located in eastern Pakistan •city in the Punjab province •in the center of the Punjab Plains
Rawalpindi: •located in eastern Pakistan •city in the Punjab province• in the center of the Punjab Plains

Multan: •located in eastern Pakistan •city in the Punjab province •in the southwestern part of the Punjab Plains
Hyderabad: •located in southeastern Pakistan •city in the Sindh province •chief port on the Indus River
Gujranwala: •located in eastern Pakistan •city in the Punjab province •in the northern part of the Punjab Plains
Peshawar: •located in northwestern Pakistan •city in the North-West Frontier province
Islamabad: •located in northern Pakistan •city in the Punjab province •capital city of Pakistan (832,000 people)

ENVIRONMENTAL/ECONOMIC:
Climate: mostly desert; temperate in the northwest; arctic at high elevations
Natural Resources: land, natural gas, petroleum, coal, iron ore
Agricultural Products: cotton, wheat, rice, sugarcane, milk
Major Exports: textiles, rice, leather, sporting goods, carpets and rugs
Natural Hazards: earthquakes, floods

PALAU
Country Name: Republic of Palau
Continent: Australia/Oceania
Area: 189 sq mi / 489 sq km
Population: 21,000
Population Density: 111 people per sq mi / 43 people per sq km
Capital: Melekeok

PHYSICAL:
Highest Point: Mount Ngerchelchuus 794 ft / 242 m
Lowest Point: Pacific Ocean 0 ft / 0 m (sea level)
Oceans:
   Pacific: •forms the eastern coast of Palau
Seas:
   Philippine: •forms the western coast of Palau •feeds into the Pacific Ocean
Islands:
   Babelthuap: •largest island in Palau •bordered by the Pacific Ocean and Philippine Sea

POLITICAL:
Independence: October 1, 1994 (from United States administration)
Former Name: Palau District (part of the Trust Territory of Pacific Islands)
Administrative Divisions: Aimeliik, Airai, Angaur, Hatohobei, Kayangel, Koror, Melekeok, Ngaraard, Ngarchelong, Ngardmau, Ngatpang, Ngchesar, Ngeremlengui, Ngiwal, Peleliu, Sonsoral (16 states)
Ethnic/Racial Groups: Palauan, Asian, white
Religions: Christianity (Roman Catholic, Protestant), Modekngei (indigenous)

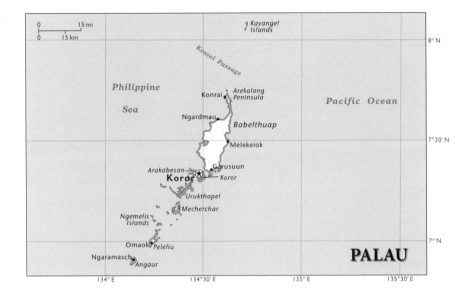

Languages: English, Palauan, Japanese, and three additional local languages
Currency: U.S. dollar
Current President: Johnson Toribiong
Cities (capital, largest, or with at least a million people):
   Koror: •located on a small island south of Babelthuap •city in the Koror state •former capital of Palau (government offices have been moved to the state of Melekeok) •most populated city in Palau (14,000 people) •chief port on the Philippine Sea and Pacific Ocean

ENVIRONMENTAL/ECONOMIC:
Climate: tropical; hot and humid with rainy/dry seasons
Natural Resources: forests, minerals, marine products, deep-seabed minerals
Agricultural Products: coconuts, copra, cassava (tapioca), sweet potatoes
Major Exports: shellfish, tuna, copra, garments
Natural Hazards: typhoons

PANAMA
Country Name: Republic of Panama
Continent: North America
Area: 29,157 sq mi / 75,517 sq km
Population: 3,571,000
Population Density: 122 people per sq mi / 47 people per sq km
Capital: Panama City

PHYSICAL:
Highest Point: Volcan Baru 11,400 ft / 3,475 m
Lowest Point: Pacific Ocean 0 ft / 0 m (sea level)
Mountain Ranges:
   Cordillera de Talamanca: •covers much of western Panama •extends into Costa Rica
Peaks (minimum elevation of 8,000 ft / 2,400 m):
   Volcan Baru: •located in the Cordillera de Talamanca •highest point in Panama
Oceans:
   Pacific: •forms much of the southern coast of Panama
Seas:
   Caribbean: •forms most of the northern coast of Panama •feeds into the
      Atlantic Ocean
Gulfs:
   Gulf of Panama: •forms much of the southern coast of Panama •feeds into the
      Pacific Ocean
Canals:
   Panama: •connects the Gulf of Panama and Caribbean Sea
Lakes:
   Lake Bayano: •located in eastern Panama
   Lake Gatun: •located in central Panama •near the end of the Panama Canal

POLITICAL:
Independence: November 3, 1903 (from Colombia)
Bordering Countries: Costa Rica, Colombia (2)
Administrative Divisions: Bocas del Toro, Chiriqui, Cocle, Colon, Darien,
   Herrera, Los Santos, Panama, Veraguas (9 provinces) •San Blas (1 territory)

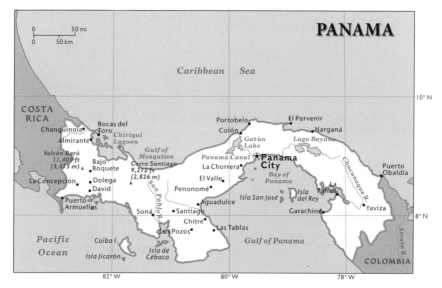

Ethnic/Racial Groups: mestizo, Amerindian–West Indian, white, Amerindian
Religion: Christianity (Roman Catholic, Protestant)
Languages: Spanish, English
Currency: balboa, U.S. dollar
Current President: Ricardo Martinelli Berrocal
Cities (capital, largest, or with at least a million people):
  Panama City: ●located in central Panama ●city in the Panama province ●capital of Panama ●most populated city in Panama (1,346,000 people)

ENVIRONMENTAL/ECONOMIC:
Climate: tropical marine; hot and humid with rainy/dry seasons
Natural Resources: copper, mahogany forests, shrimp, hydropower
Agricultural Products: bananas, rice, corn, coffee, livestock
Major Exports: bananas, shrimp, sugar, coffee, clothing
Natural Hazards: severe storms, forest fires

PAPUA NEW GUINEA
Country Name: Independent State of Papua New Guinea
Continent: Australia/Oceania
Area: 178,703 sq mi / 462,840 sq km
Population: 6,888,000
Population Density: 39 people per sq mi / 15 people per sq km
Capital: Port Moresby

PHYSICAL:
Highest Point: Mount Wilhelm 14,793 ft / 4,509 m
Lowest Point: Pacific Ocean 0 ft / 0 m
Mountain Ranges:
  Bismarck: ●located on the island of New Guinea
  Maoke: ●located on the island of New Guinea ●extend into Indonesia
  Owen Stanley: ●located on the island of New Guinea
Peaks (minimum elevation of 8,000 ft / 2,400 m):
  Mount Wilhelm: ●located in the Bismarck Range ●highest point in Australia/Oceania
Oceans:
  Pacific: ●forms much of the northern coast of Papua New Guinea
Seas:
  Bismarck: ●forms much of the northern coast of Papua New Guinea ●feeds into the Pacific Ocean
  Coral: ●forms much of the southern coast of Papua New Guinea ●feeds into the Pacific Ocean
  Solomon: ●forms much of the eastern coast of Papua New Guinea ●feeds into the Pacific Ocean

Gulfs:
  Gulf of Papua: •forms part of the southern coast of Papua New Guinea •feeds into the Coral Sea
Straits:
  Torres: •forms part of the southwestern coast of Papua New Guinea •separates the island of New Guinea from Australia •connects the Coral and Arafura Seas
Islands:
  Bougainville: •island bordered by the Pacific Ocean and Solomon Sea
  New Britain: •second largest island in Papua New Guinea •bordered by the Bismarck and Solomon Seas
  New Guinea: •largest island in Papua New Guinea •bordered by the Bismarck Sea, Solomon Sea, Coral Sea, Gulf of Papua, and Torres Strait •western half belongs to Indonesia
  New Ireland: •island bordered by the Pacific Ocean and Bismarck Sea

POLITICAL:
Independence: September 16, 1975 (from Australian administration)
Former Name: Territory of Papua New Guinea
Bordering Country: Indonesia (1)
Administrative Divisions: Bougainville, Central, Chimbu, Eastern Highlands, East New Britain, East Sepik, Enga, Gulf, Madang, Manus, Milne Bay, Morobe, National Capital, New Ireland, Northern, Sandaun, Southern Highlands, Western, Western Highlands, West New Britain (20 provinces)
Ethnic/Racial Groups: Melanesian, Papuan, Negrito, Micronesian, Polynesian
Religions: Christianity (Protestant, Roman Catholic), indigenous beliefs

Languages: 715 indigenous languages
Currency: kina
Current Governor: Michael Ogio
Cities (capital, largest, or with at least a million people):
  Port Moresby: •located on the island of New Guinea •city in the National Capital province •capital of Papua New Guinea •most populated city in Papua New Guinea (314,000 people) •chief port on the Gulf of Papua

ENVIRONMENTAL/ECONOMIC:
Climate: tropical; monsoons with little temperature variation
Natural Resources: gold, copper, silver, natural gas, timber
Agricultural Products: coffee, cacao, coconuts, palm kernels, poultry
Major Exports: oil, gold, copper ore, logs
Natural Hazards: volcanoes, earthquakes, mudslides, tsunamis

PARAGUAY
Country Name: Republic of Paraguay
Continent: South America
Area: 157,048 sq mi / 406,752 sq km
Population: 6,568,000
Population Density: 42 people per sq mi / 16 people per sq km
Capital: Asuncion

PHYSICAL:
Highest Point: Cerro Pero 2,762 ft / 842 m
Lowest Point: Paraguay/Parana River junction 151 ft / 46 m
Grasslands/Prairies:
  Gran Chaco: •covers much of Paraguay •extends into Argentina
Swamps:
  Pantanal: •located in northeastern Paraguay •extends into Bolivia and Brazil •largest swamp in the world
Rivers:
  Paraguay: •tributary of the Parana River •forms part of Paraguay's border with Brazil and Argentina •extends into Brazil
  Parana: •forms part of Paraguay's border with Argentina and Brazil •extends into Brazil
  Pilcomayo: •tributary of the Paraguay River •forms part of the Paraguay-Argentina border •extends into Bolivia
Lakes:
  Itaipu Reservoir: •located in southeastern Paraguay •on the Paraguay-Brazil border •fed by the Parana River •formed by the Itaipu Dam
Dams:
  Itaipu: •located in southeastern Paraguay •on the Paraguay-Brazil border •crosses the Parana River •forms Itaipu Reservoir

PARAGUAY

POLITICAL:
Independence: May 14, 1811 (from Spain)
Bordering Countries: Argentina, Brazil, Bolivia (3)
Administrative Divisions: Alto Paraguay, Alto Parana, Amambay, Boqueron, Caaguazu, Caazapa, Canindeyu, Central, Concepcion, Cordillera, Guaira, Itapua, Misiones, Neembucu, Paraguari, Presidente Hayes, San Pedro (17 departments) •Asuncion (1 capital city)
Ethnic/Racial Groups: mestizo
Religion: Christianity (primarily Roman Catholic)
Languages: Spanish, Guarani
Currency: guarani
Current President: Fernando Armindo Lugo Mendez
Cities (capital, largest, or with at least a million people):
  Asuncion: •located in southwestern Paraguay •city in the Asuncion capital city division •capital of Paraguay •most populated city in Paraguay (1,977,000 people) •chief port on the Paraguay River

ENVIRONMENTAL/ECONOMIC:
Climate: subtropical to temperate; rainy in the east; semi-arid in the far west
Natural Resources: hydropower, timber, iron ore, manganese, limestone
Agricultural Products: cotton, sugarcane, soybeans, corn, beef
Major Exports: soybeans, feed, cotton, meat, edible oils
Natural Hazards: floods

PERU

Country Name: Republic of Peru
Continent: South America
Area: 496,224 sq mi / 1,285,216 sq km
Population: 29,400,000
Population Density: 59 people per sq mi / 23 people per sq km
Capital: Lima

PHYSICAL:

Highest Point: Mt. Huascaran 22,205 ft / 6,768 m
Lowest Point: Pacific Ocean 0 ft / 0 m (sea level)
Mountain Ranges:
  Andes: • cover much of Peru • extend into Ecuador, Bolivia, and Chile
Peaks (minimum elevation of 8,000 ft / 2,400 m):
  Mt. Huascaran: • located in the Andes • highest point in Peru
Plateau Regions:
  Altiplano: • located in southeastern Peru • extends into Bolivia • basin in the Andes
Rain Forests:
  Amazon: • covers much of northern and eastern Peru • extends into Ecuador, Colombia, Brazil, and Bolivia • largest rain forest in the world
Deserts:
  Atacama: • located in southeastern Peru • extends into Chile • driest place in the world
  Sechura: • located in northwestern Peru

Oceans:
    Pacific: •forms most of the western and southern coasts of Peru
Gulfs:
    Gulf of Guayaquil: •forms part of the northwestern coast of Peru •feeds into the Pacific Ocean
Lakes:
    Lake Titicaca: •located in southeastern Peru •on the Altiplano near the slopes of the Andes •extends into Bolivia •largest lake in South America
Rivers:
    Amazon: •has its source in the Andes •forms part of the Peru-Colombia border •extends into Brazil •longest river in South America
    Maranon: •tributary of the Amazon River •has its source in the Andes
    Pastaza: •tributary of the Maranon River •extends into Ecuador
    Purus: •has its source in the Andes •extends into Brazil
    Putumayo: •forms part of the Peru-Colombia border •extends into Colombia
    Ucayali: •tributary of the Amazon River •has its source in the Andes
    Yurua: •has its source in the Andes •extends into Brazil

POLITICAL:
Independence: July 28, 1821 (from Spain)
Bordering Countries: Brazil, Colombia, Ecuador, Bolivia, Chile (5)
Administrative Divisions: Amazonas, Ancash, Apurimac, Arequipa, Ayacucho, Cajamarca, Cusco, Huancavelica, Huanuco, Ica, Junin, La Libertad, Lambayeque, Lima, Loreto, Madre de Dios, Moquegua, Pasco, Piura, Puno, San Martin, Tacna, Tumbes, Ucayali (24 departments) •Callao (1 constitutional province)
Ethnic/Racial Groups: Amerindian, mestizo, white
Religion: Christianity (primarily Roman Catholic)
Languages: Spanish, Quechua, Aymara
Currency: nuevo sol
Current President: Ollanta Humala Tasso
Cities (capital, largest, or with at least a million people):
    Lima: •located in western Peru •in the Lima department and the Callao constitutional province •capital of Peru •most populated city in Peru (8,769,000 people) •on the edge of the Andes

ENVIRONMENTAL/ECONOMIC:
Climate: tropical in the east; dry desert in the west; temperate to frigid in the Andes
Natural Resources: copper, silver, gold, petroleum, timber
Agricultural Products: coffee, cotton, sugarcane, rice, poultry
Major Exports: fish and fish products, gold, copper, zinc, crude petroleum and by-products
Natural Hazards: earthquakes, tsunamis, floods, landslides, volcanoes

PHILIPPINES

Country Name: Republic of the Philippines
Continent: Asia
Area: 115,831 sq mi / 300,000 sq km
Population: 95,739,000
Population Density: 826 people per sq mi / 319 people per sq km
Capital: Manila

PHYSICAL:

Highest Point: Mount Apo 9,692 ft / 2,954 m
Lowest Point: Philippine Sea 0 ft / 0 m (sea level)
Peaks (minimum elevation of 8,000 ft / 2,400 m):
　Mount Apo: •located on the island of Mindanao •highest point in the Philippines
Seas:
　Bohol: •forms part of the inner coast of the Philippines •feeds into the Sulu and Philippine Seas
　Celebes: •forms most of the southern coast of Mindanao •feeds into the Philippine Sea
　Philippine: •forms most of the eastern coast of the Philippines •feeds into the Pacific Ocean
　Sibuyan: •forms part of the inner coast of the Philippines •feeds into the Sulu and Philippine Seas
　South China: •forms most of the western coast of the Philippines •feeds into the Philippine and Java Seas
　Sulu: •forms part of the inner coast of the Philippines •feeds into the South China and Celebes Seas

    Visayan: •forms part of the inner coast of the Philippines •feeds into the Sulu and Philippine Seas
Gulfs:
    Davao: •forms part of the southeastern coast of Mindanao •feeds into the Philippine Sea
    Moro: •forms part of the southern coast of Mindanao •feeds into the Celebes Sea
Bays:
    Manila: •forms part of the southwestern coast of Luzon •feeds into the South China Sea
Straits:
    Luzon: •forms much of the northern coast of Luzon •separates the Philippines from the island of Taiwan •connects the South China and Philippine Seas
Islands:
    Cebu: •island between Negros and Leyte •bordered by the Visayan and Bohol Seas
    Leyte: •island between Samar and Cebu •bordered by the Visayan and Bohol Seas
    Luzon. • largest island in the Philippines •bordered by the South China, Sibuyan, and Philippine Seas
    Mindanao: •second largest island in the Philippines •bordered by the Sulu Sea, Bohol Sea, Philippine Sea, Celebes Sea, Moro Gulf, and Davao Gulf
    Mindoro: •island off the southern coast of Luzon •bordered by the South China, Sulu, and Sibuyan Seas
    Negros: •island between Panay and Cebu •bordered by the Sulu, Visayan, and Bohol Seas
    Palawan: •island between the South China and Sulu Seas
    Panay: •island between Negros and Mindoro •bordered by the Sulu, Sibuyan, and Visayan Seas
    Samar: •island off the southeastern coast of the Philippines •bordered by the Philippine Sea

POLITICAL:

Independence: July 4, 1946 (from the United States)

Administrative Divisions: Abra, Agusan del Norte, Agusan del Sur, Aklan, Albay, Antique, Apayao, Aurora, Basilan, Bataan, Batanes, Batangas, Benguet, Biliran, Bohol, Bukidnon, Bulacan, Cagayan, Camarines Norte, Camarines Sur, Camiguin, Capiz, Catanduanes, Cavite, Cebu, Compostela, Davao del Norte, Davao del Sur, Davao Oriental, Dinagat Islands, Eastern Samar, Guimaras, Ifugao, Ilocos Norte, Iloco Sur, Iloilo, Isabela, Kalinga, Laguna, Lanao del Norte, Lanao del Sur, La Union, Leyte, Maguindanao, Marinduque, Masbate, Mindoro Occidental, Mindoro Oriental, Misamis Occidental, Misamis Oriental, Mountain, Negros Occidental, Negros Oriental, North Cotabato, Northern Samar, Nueva Ecija, Nueva Vizcaya, Palawan, Pampanga, Pangasinan, Quezon, Quirino, Rizal, Romblon, Samar, Sarangani, Siquijor, Sorsogon, South Cotabato, Southern Leyte, Sultan Kudarat, Sulu, Surigao del Norte, Surigao del Sur, Tarlac, Tawi-Tawi, Zambales, Zamboanga del Norte, Zamboanga del Sur, Zamboanga Sibugay (80 provinces) •Alaminos,

Angeles, Antipolo, Bacolod, Bago, Baguio, Bais, Balanga, Batac, Batangas, Bayawan, Bislig, Butuan, Cabadbaran, Cabanatuan, Cadiz, Cagayan de Oro, Calamba, Calapan, Calbayog, Candon, Canlaon, Cauayan, Cavite, Cebu, Cotabato, Dagupan, Danao, Dapitan, Davao, Digos, Dipolog, Dumaguete, Escalante, Gapan, General Santos, Gingoog, Himamaylan, Iligan, Iloilo, Iriga, Isabela, Kabankalan, Kalookan, Kidapawan, Koronadal, La Carlota, Laoag, Lapu-Lapu, Las Pinas, Legazpi, Ligao, Lipa, Lucena, Maasin, Makati, Malabon, Malaybalay, Malolos, Mandaluyong, Mandaue, Manila, Marawi, Marikina, Masbate, Mati, Meycauayan, Munoz, Muntinlupa, Naga, Navotas, Olongapo, Ormoc, Oroquieta, Ozamis, Pagadian, Palayan, Panabo, Paranaque, Pasay, Pasig, Passi, Puerto Princesa, Quezon, Roxas, Sagay, Samal, San Carlos (in Negros Occidental), San Carlos (in Pangasinan), San Fernando (in La Union), San Fernando (in Pampanga), San Jose, San Jose del Monte, San Juan, San Pablo, Santa Rosa, Santiago, Silay, Sipalay, Sorsogon, Surigao, Tabaco, Tacloban, Tacurong, Tagaytay, Tagbilaran, Taguig, Tagum, Talisay (in Cebu), Talisay (in Negros Oriental), Tanauan, Tangub, Tanjay, Tarlac, Toledo, Trece Martires, Tuguegarao, Urdaneta, Valencia, Valenzuela, Victorias, Vigan, Zamboanga (123 chartered cities)
Ethnic/Racial Groups: Christian Malay, Muslim Malay, Chinese
Religions: Christianity (Roman Catholic, Protestant), Islam, Buddhism
Languages: Filipino (based on Tagalog), English, 8 major dialects
Currency: Philippine peso
Current President: Benigno Aquino
Cities (capital, largest, or with at least a million people):
  Manila: •located on the island of Luzon •city in the Manila chartered city •capital of the Philippines •most populated city in the Philippines (11,449,000 people) •chief port on Manila Bay
  Davao: •most populated city on the island of Mindanao •chief cargo transport hub

ENVIRONMENTAL/ECONOMIC:
Climate: tropical marine with monsoons
Natural Resources: timber, petroleum, nickel, cobalt, silver
Agricultural Products: rice, coconuts, corn, sugarcane, pork
Major Exports: electronic equipment, machinery and transport equipment, garments, coconut products
Natural Hazards: typhoons, cyclones, landslides, volcanoes, earthquakes, tsunamis

POLAND
Country Name: Republic of Poland
Continent: Europe
Area: 120,728 sq mi / 312,685 sq km
Population: 38,222,000
Population Density: 317 people per sq mi / 122 people per sq km
Capital: Warsaw

PHYSICAL:
Highest Point: Rysy 8,199 ft / 2,499 m
Lowest Point: near Raczki Elblaskie 7 ft / 2 m below sea level
Mountain Ranges:
    Carpathian: •cover much of southern Poland •extend into Czech Republic, Slovakia, and Ukraine
    Sudeten: •located in southwestern Poland •extend into Czech Republic
Peaks (minimum elevation of 8,000 ft / 2,400 m):
    Rysy: •located in the Carpathian Mountains •on the Poland-Czech Republic border •highest point in Poland
Seas:
    Baltic: •forms part of the northern coast of Poland •feeds into the Kattegat
Gulfs:
    Gulf of Gdansk: •forms the northeastern coast of Poland •feeds into the Baltic Sea
Bays:
    Pomeranian: •forms part of the northwestern coast of Poland •feeds into the Baltic Sea
Rivers:
    Odra (Oder): •has its mouth in the Stettiner Haff •forms part of the Poland-Germany border •extends into Czech Republic
    Wista (Vistula): •has its mouth in the Gulf of Gdansk •has its source in the Carpathian Mountains
    Warta: •tributary of the Oder River •has its source in southern Poland
Lagoons:
    Stettiner Haff: •feeds into the Baltic Sea •on the Poland-Germany border
Regions:
    Beskids: •mountainous region that covers much of southern Poland •includes part of the Carpathian Mountains

POLITICAL:
Independence: November 11, 1918 (proclaimed an independent republic)
Bordering Countries: Czech Republic, Ukraine, Germany, Slovakia, Belarus, Russia (Kaliningrad province), Lithuania (7)
Regions:
    Galicia: •region in southeastern Poland •includes part of the Carpathian Mountains •extends into Ukraine
    Masuria: •region in northeastern Poland
    Pomerania: •region that covers much of northern Poland •south of the Baltic Sea •extends into Germany
    Silesia: •region in southwestern Poland •north of the Sudeten Mountains
Administrative Divisions: Dolnoslaskie, Kujawsko-Pomorskie, Lodzkie, Lubelskie, Lubuskie, Malopolskie, Mazowieckie, Opolskie, Podkarpackie, Podlaskie, Pomorskie, Slaskie, Swietokrzyskie, Warminsko-Mazurskie, Wielkopolskie, Zachodniopomorskie (16 provinces)
Ethnic/Racial Groups: predominantly Polish
Religion: Christianity (primarily Roman Catholic)

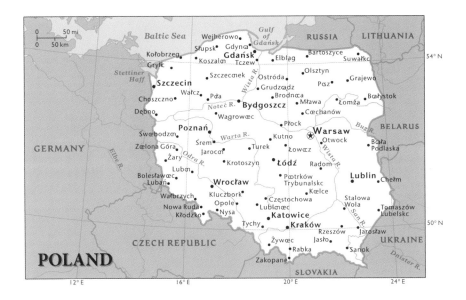

Language: Polish
Currency: zloty
Current President: Bronislaw Komorowski
Cities (capital, largest, or with at least a million people):
　Warsaw:　•located in eastern Poland　•city in the Mazowieckie province　•capital of Poland　•most populated city in Poland (1,710,000 people)　•chief port on the Wista (Vistula) River
　Lodz:　•located in central Poland　•city in the Lodzkie province

ENVIRONMENTAL/ECONOMIC:

Climate: temperate wet; cold winters/mild summers
Natural Resources: coal, sulfur, copper, natural gas, silver
Agricultural Products: potatoes, fruits, vegetables, wheat, poultry
Major Exports: machinery and transport equipment, intermediate manufactured goods
Natural Hazards: floods

PORTUGAL

Country Name: Portuguese Republic
Continent: Europe
Area: 35,655 sq mi / 92,345 sq km
Population: 10,653,000
Population Density: 299 people per sq mi / 115 people per sq km
Capital: Lisbon

PHYSICAL:

Highest Point: Ponta do Pico (in the Azores) 7,713 ft / 2,351 m
Lowest Point: Atlantic Ocean 0 ft / 0 m (sea level)
Oceans:
   Atlantic: •forms the western coast of Portugal
Gulfs:
   Gulf of Cadiz: •forms the southeastern coast of Portugal •feeds into the Atlantic Ocean
Rivers:
   Guadiana: •has its mouth in the Gulf of Cadiz •forms part of the Portugal-Spain border •extends into Spain
   Tagus: •has its mouth in the Atlantic Ocean •forms part of the Portugal-Spain border •extends into Spain
Islands:
   Azores: •island group in the middle of the Atlantic Ocean
   Madeira: •island group in the eastern Atlantic Ocean •off the coast of Morocco

POLITICAL:

Independence: 1140 (established as an independent kingdom; declared an independent republic October 5, 1910)
Bordering Country: Spain (1)
Administrative Divisions: Aveiro, Beja, Braga, Braganca, Castelo Branco, Coimbra, Evora, Faro, Guarda, Leiria, Lisboa, Portalegre, Porto, Santarem, Setubal, Viana do Castelo, Vila Real, Viseu (18 districts) •Azores, Madeira (2 autonomous regions)
Ethnic/Racial Groups: predominantly Portuguese
Religion: Christianity (primarily Roman Catholic)
Languages: Portuguese, Mirandese
Currency: euro
Current President: Anibal Cavaco Silva
Cities (capital, largest, or with at least a million people):
  Lisbon: •located in southwestern Portugal •city in the Lisboa district •capital of Portugal •most populated city in Portugal (2,808,000 people) •chief port on the Tagus River and Atlantic Ocean

ENVIRONMENTAL/ECONOMIC:

Climate: temperate marine, cool and rainy in the north; warmer and drier in the south
Natural Resources: fish, forests (cork), iron ore, copper, zinc
Agricultural Products: grain, potatoes, olives, grapes, sheep
Major Exports: clothing and footwear, machinery, chemicals, cork and paper products, hides
Natural Hazards: earthquakes (in the Azores)

QATAR

Country Name: State of Qatar
Continent: Asia
Area: 4,448 sq mi / 11,521 sq km
Population: 1,732,000
Population Density: 389 people per sq mi / 150 people per sq km
Capital: Doha

PHYSICAL:

Highest Point: Tuwayyir al Hamir 338 ft / 103 m
Lowest Point: Persian Gulf 0 ft / 0 m (sea level)
Deserts:
  Rub al Khali: •covers most of the country •extends into Saudi Arabia
Gulfs:
  Gulf of Bahrain: •forms most of the western coast of Qatar •feeds into the Persian Gulf
  Persian: •forms the northern and eastern coasts of Qatar •feeds into the Gulf of Oman

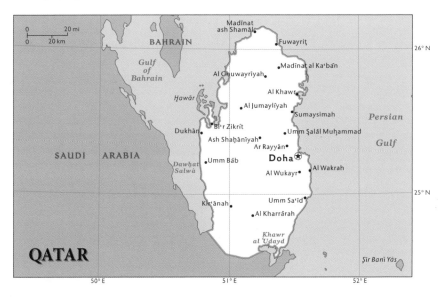

POLITICAL:
Independence: September 3, 1971 (from the United Kingdom)
Bordering Country: Saudi Arabia (1)
Administrative Divisions: Ad Dawhah, Al Khawr wa adh Dhakhirah, Al Wakrah, Ar Rayyan, Ash Shamal, Az Zaayin, Umm Salal (7 municipalities)
Ethnic Groups: Arab, Pakistani, Indian, Iranian
Religion: primarily Islam
Languages: Arabic, English
Currency: Qatari rial
Current Prime Minister: Hamad bin Jasmin bin Jaber al-Thani
Cities (capital, largest, or with at least a million people):
  Doha: •located in eastern Qatar •city in the Ad Dawhah municipality •capital of Qatar •most populated city in Qatar (427,000 people) •chief port on the Persian Gulf

ENVIRONMENTAL/ECONOMIC:
Climate: arid with mild winters/hot and humid summers
Natural Resources: petroleum, natural gas, fish
Agricultural Products: fruits, vegetables, poultry, dairy products, beef
Major Exports: petroleum products, fertilizers, steel
Natural Hazards: haze, dust storms, sandstorms

COUNTRIES

ROMANIA

Country Name: Romania
Continent: Europe
Area: 92,043 sq mi / 238,391 sq km
Population: 21,408,000
Population Density: 233 people per sq mi / 90 people per sq km
Capital: Bucharest

PHYSICAL:

Highest Point: Moldoveanu 8,346 ft / 2,544 m
Lowest Point: Black Sea 0 ft / 0 m (sea level)
Mountain Ranges:
  Carpathian: •cover much of northern Romania •extend into Ukraine
  Transylvanian Alps: •cover much of central Romania
Peaks (with minimum elevation of 8,000 ft / 2,400 m):
  Moldoveanu: •located in the Transylvanian Alps •highest point in Romania
Grasslands/Prairies:
  Black Sea Lowland: •located in southeastern Romania •extends into Moldova and Ukraine
  Great Alfold: •covers much of western Romania •extends into Ukraine, Hungary, and Serbia and Montenegro
Seas:
  Black: •forms the southeastern coast of Romania •feeds into the Sea of Marmara
Rivers:
  Danube: •has its mouth in the Black Sea •forms most of the Romania-Bulgaria border •forms part of Romania's border with Serbia and Ukraine •extends into Hungary and Serbia
  Prut: •tributary of the Danube River •forms all of the Romania-Moldova border •forms part of the Romania-Ukraine border •extends into Ukraine
Deltas:
  Delta of the Danube: •mouth of the Danube River •feeds into the Black Sea •borders Ukraine

POLITICAL:

Independence: 1878 (independence from Turkey officially recognized)
Bordering Countries: Bulgaria, Ukraine, Serbia, Moldova, Hungary (5)

PAGE 278 | THE ULTIMATE FACT BOOK

Regions:
Dobruja: •region in southeastern Romania •west of the Black Sea •east and south of the Danube River •extends into Bulgaria
Moldavia: •region in eastern Romania •east of the Carpathian Mountains •west of the Prut River
Transylvania: •region in northwestern Romania •west of the Carpathian Mountains •north of the Transylvanian Alps
Walachia: •region in southern Romania •south of the Transylvanian Alps •north of the Danube River
Administrative Divisions: Alba, Arad, Arges, Bacau, Bihor, Bistrita-Nasaud, Botosani, Braila, Brasov, Buzau, Calarasi, Caras-Severin, Cluj, Constanta, Covasna, Dimbovita, Dolj, Galati, Giurgiu, Gorj, Harghita, Hunedoara, Ialomita, Iasi, Ilfov, Maramures, Mehedinti, Mures, Neamt, Olt, Prahova, Salaj, Satu Mare, Sibiu, Suceava, Teleorman, Timis, Tulcea, Vaslui, Valcea, Vrancea (41 counties) •Bucuresti (1 municipality)
Ethnic/Racial Groups: Romanian, Hungarian, Roma
Religion: Christianity (Eastern Orthodox, Protestant, Roman Catholic)
Languages: Romanian, Hungarian, German
Currency: leu
Current President: Traian Basescu
Cities (capital, largest, or with at least a million people):
Bucharest: •located in southern Romania •city in the Bucuresti municipality •capital of Romania •most populated city in Romania (1,933,000 people)

ENVIRONMENTAL/ECONOMIC:
Climate: temperate with cold and snowy winters/sunny summers with occasional heavy rains
Natural Resources: petroleum, timber, natural gas, coal, iron ore
Agricultural Products: wheat, corn, barley, sugar beets, eggs
Major Exports: textiles and footwear, metals and metal products, machinery and equipment, minerals and fuels
Natural Hazards: earthquakes, landslides

RUSSIA
Country Name: Russian Federation
Continent: Asia/Europe (most of the land area is in Asia; the capital city and most of the population is in Europe)
Area: 6,592,850 sq mi / 17,075,400 sq km (largest country in the world)
Population: 142,847,000
Population Density: 22 people per sq mi / 8 people per sq km
Capital: Moscow

PHYSICAL:
Highest Point: Mount Elbrus 18,510 ft / 5,642 m
Lowest Point: Caspian Sea 92 ft / 28 m below sea level
Mountain Ranges:
    Altay: •located in southern Russia •extend into Kazakhstan, China, and Mongolia
    Caucasus: •located in southwestern Russia •extend into Georgia and Azerbaijan
    Cherskiy Range: •located in northeastern Russia
    Eastern Sayan: •located in southern Russia •extend into Mongolia
    Kolyma Range: •located in northeastern Russia
    Ural: •located in western Russia •form part of what is sometimes recognized as the boundary between the continents of Asia and Europe
    Western Sayan: •located in southern Russia
    Yablonovyy Range: •located in southern Russia
Peaks (with minimum elevation of 8,000 ft / 2,400 m):
    Mount Elbrus: •located in the Caucasus •highest point in Russia •highest point in Europe
Grasslands/Prairies:
    North Siberian Lowland: •located in northern Russia
    West Siberian Plain: •covers much of central Russia
Plateau Regions:
    Aldan Plateau: •located in eastern Russia
    Central Siberian Plateau: •covers much of central Russia
Depressions:
    Caspian Depression: •located in southwestern Russia •extends into Kazakhstan
    Caspian Sea: •located in southwestern Russia •lowest point in Europe

Oceans:
 Arctic: • forms much of the northern coast of Russia
 Pacific: • forms much of the eastern coast of Russia
Seas:
 Barents: • forms part of the northwestern coast of Russia • feeds into the Arctic Ocean
 Bering: • forms part of the northeastern coast of Russia • feeds into the Pacific Ocean
 Black: • forms part of the southwestern coast of Russia • feeds into the Sea of Marmara
 Chukchi: • forms part of the northeastern coast of Russia • feeds into the Arctic Ocean
 East Siberian: • forms part of the northeastern coast of Russia • feeds into the Arctic Ocean
 Kara: • forms part of the northern coast of Russia • feeds into the Arctic Ocean
 Laptev: • forms part of the northern coast of Russia • feeds into the Arctic Ocean
 Sea of Azov: • forms part of the southwestern coast of Russia • feeds into the Black Sea
 Sea of Japan: • forms part of the southeastern coast of Russia • feeds into the Sea of Okhotsk, Pacific Ocean, and East China Sea
 Sea of Okhotsk: • forms part of the eastern coast of Russia • feeds into the Pacific Ocean
 White: • forms part of the northwestern coast of Russia • feeds into the Barents Sea
Gulfs:
 Gulf of Anadyr: • forms part of the northeastern coast of Russia • feeds into the Bering Sea
 Gulf of Finland: • forms part of the western coast of Russia • feeds into the Baltic Sea
 Gulf of Ob: • forms part of the northern coast of Russia • feeds into the Kara Sea
 Shelikhov: • forms part of the eastern coast of Russia • feeds into the Sea of Okhotsk

Straits:
   Bering: •forms part of the northeastern coast of Russia •separates Russia from the United States •connects the Chukchi and Bering Seas
   La Perouse: •forms part of the southeastern coast of Sakhalin Island •separates Sakhalin Island from Japan •connects the Sea of Japan and Sea of Okhotsk
   Tatar: •forms part of the southeastern coast of Russia •separates Sakhalin Island from mainland Russia •connects the Sea of Okhotsk and Sea of Japan

Lakes:
   Caspian Sea: •located in southwestern Russia •extends into Azerbaijan and Kazakhstan •largest lake in the world •lowest point in Europe
   Lake Baikal: • located in southern Russia •on the edge of the Eastern Sayan Mountains •largest lake entirely in Asia
   Lake Ladoga: •located in northwestern Russia •largest lake entirely in Europe
   Lake Onega: •located in northwestern Russia
   Lake Peipus: •located in western Russia •on the Russia-Estonia border
   Lake Pskov: •located in western Russia •on the Russia-Estonia border
   Rybinsk Reservoir: •located in western Russia •fed by the Volga River

Rivers:
   Aldan: •tributary of the Lena River •has its source in the Aldan Plateau
   Amur: •has its mouth in the Tatar Strait •forms part of the Russia-China border
   Angara: •tributary of the Yenisey River
   Dnieper: •has its source in western Russia •extends into Belarus and Ukraine
   Don: •has its mouth in the Sea of Azov •has its source in western Russia
   Donets: •tributary of the Don River •has its source in southwestern Russia •extends into Ukraine
   Irtysh (Ertis): • tributary of the Ob •extends into Kazakhstan and China
   Kolyma: •has its mouth in the East Siberian Sea •has its source in the Cherskiy Range
   Lena: •has its mouth in the Laptev Sea •has its source in southern Russia
   Ob: • has its mouth in the Gulf of Ob • has its source in the Altay Mountains
   Pechora: • has its mouth in the Barents Sea • has its source in the Ural Mountains
   Volga: •has its mouth in the Caspian Sea •has its source in western Russia •flows through the Rybinsk Reservoir •longest river in Europe
   Western Dvina: • has its source in western Russia •extends into Belarus and Latvia
   Yenisey: •has its mouth in the Kara Sea

Deltas:
   Volga River Delta: •mouth of the Volga River •feeds into the Caspian Sea

Peninsulas:
   Chukchi: •bordered by the Chukchi Sea, Bering Sea, Gulf of Anadyr, and Bering Strait
   Kamchatka: •bordered by the Pacific Ocean, Sea of Okhotsk, Bering Sea, and Shelikhov Gulf
   Kola: •bordered by the White and Barents Seas
   Taymyr: •bordered by the Kara and Laptev Seas

Islands:
  Franz Josef Land: •island group between the Arctic Ocean and Barents Sea
  Kuril: • island chain between the Sea of Okhotsk and Pacific Ocean
  New Siberian: •island group off the coast of northeastern Russia •bordered by the Arctic Ocean, Laptev Sea and East Siberia Sea
  Novaya Zemlya: •island group off the northwestern coast of Russia •bordered by the Barents and Kara Seas
  Sakhalin: •island separated from the mainland by the Tatar Strait •bordered by the Sea of Okhotsk, Tatar Strait, and La Perouse Strait
  Severnaya Zemlya: •island group off the coast of the Taymyr Peninsula •bordered by the Arctic Ocean, Kara Sea, and Laptev Sea

Regions:
  Siberia: •covers much of northern and central Russia •south of the Kara Sea, Laptev Sea, East Siberian Sea, Chukchi Sea, and Gulf of Ob •west of the Bering Sea, Shelikhov Gulf, and Pacific Ocean •east of the Ural Mountains •north of the Altay Mountains, Western Sayan Mountains, Eastern Sayan Mountains, Aldan Plateau, Sea of Okhotsk, and Shelikhov Gulf •includes the Cherskiy Range, Kolyma Range, Central Siberian Plateau, West Siberian Plain, North Siberian Lowland, Taymyr Peninsula, Chukchi Peninsula, and Kamchatka Peninsula

POLITICAL:
Independence: August 24, 1991 (from the Soviet Union)
Former Names: Russian Empire, Russian Soviet Federated Socialist Republic
Bordering Countries: Kazakhstan, China, Mongolia, Ukraine, Finland, Belarus, Georgia, Estonia, Azerbaijan, Lithuania, Latvia, Poland, Norway, North Korea (14)
Administrative Divisions: Amur, Arkhangel'sk, Astrakhan', Belgorod, Bryansk, Chelyabinsk, Irkutsk, Ivanovo, Kaliningrad, Kaluga, Kemerovo, Kirov, Kostroma, Kurgan, Kursk, Leningrad, Lipetsk, Magadan, Moscow, Murmansk, Nizhniy Novgorod, Novgorod, Novosibirsk, Omsk, Orenburg, Orel, Penza, Pskov, Rostov, Ryazan', Sakhalin, Samara, Saratov, Smolensk, Sverdlovsk, Tambov, Tomsk, Tula, Tver', Tyumen', Ul'yanovsk, Vladimir, Volgograd, Vologda, Voronezh, Yaroslavl' (46 provinces) •Adygeya, Altay, Bashkortostan, Buryatiya, Chechnya, Chuvashiya, Dagestan, Ingushetiya, Kabardino-Balkariya, Kalmykiya, Karachayevo-Cherkesiya, Kareliya, Khakasiya, Komi, Mariy-El, Mordoviya, North Ossetia, Sakha, Tatarstan, Tyva, Udmurtiya (21 republics) •Chukotka, Khanty-Mansi, Nenets, Yamalo-Nenets (4 autonomous okrugs) •Altay, Kamchatka, Khabarovsk, Krasnodar, Krasnoyarsk, Perm', Primorskiy, Stavropol', Zabaykal'sk (9 krays) •Moscow, Saint Petersburg (2 federal cities) •Yevrey (1 autonomous oblast)
Ethnic/Racial Groups: Russian, Tatar, Ukrainian
Religions: Christianity (Russian Orthodox), Islam, other
Language: Russian
Currency: Russian ruble
Current President: Dmitriy Anatolyevich Medvedev

Cities (capital, largest, or with at least a million people):
  Moscow: •located in western Russia •city in the Moscow federal city division •capital of Russia •most populated city in Russia (10,523,000 people)
  Saint Petersburg: •located in northwestern Russia •city in the Saint Petersburg federal city division •chief port on the Gulf of Finland
  Novosibirsk: •located in southern Russia •city in the Novosibirskaya oblast
  Yekaterinburg: •located in western Russia •city in the Sverdloskaya oblast
  Nizhniy Novgorod: •located in western Russia •city in the Nizhegorodskaya oblast •chief port on the Volga River
  Samara: •located in western Russia •city in the Samarskaya oblast
  Omsk: •located in southern Russia •city in the Omskaya oblast
  Kazan: •located in western Russia •city in the Tatarstan republic
  Rostov na Donu: •located in western Russia •city in the Rostovskaya oblast
  Chelyabinsk: •located in western Russia •city in the Chelyabinskaya oblast
  Volgograd: •located in western Russia •city in the Volgogradskaya oblast
  Ufa: •located in western Russia •city in the Bashkortostan republic

ENVIRONMENTAL/ECONOMIC:
Climate: steppe in the south; humid continental in Europe; subarctic to tundra in Siberia; polar in the north; cool winters along the Black Sea; frigid winters in Siberia/warm summers in the steppes and cool along the Arctic coast
Natural Resources: oil, natural gas, coal, minerals, timber
Agricultural Products: grain, sugar beets, sunflower seeds, vegetables, beef
Major Exports: petroleum and petroleum products, natural gas, wood and wood products, metals, fur
Natural Hazards: permafrost, volcanoes, earthquakes, floods, forest fires

RWANDA
Country Name: Republic of Rwanda
Continent: Africa
Area: 10,169 sq mi / 26,338 sq km
Population: 10,943,000
Population Density: 1,076 people per sq mi / 415 people per sq km
Capital: Kigali

PHYSICAL:
Highest Point: Mount Karisimbi 14,826 ft / 4,519 m
Lowest Point: Ruzizi River 3,117 ft / 950 m
Mountain Ranges:
  Virunga: •cover most of western and central Rwanda •extend into Democratic Republic of the Congo and Burundi
Peaks (with minimum elevation of 8,000 ft / 2,400 m):
  Mount Karisimbi: •located in the Virunga Mountains •on the Rwanda-Democratic Republic of the Congo border •highest point in Rwanda

Valleys:
>Great Rift: •located in western Rwanda •extends into Democratic Republic of the Congo, Uganda, and Burundi

Lakes:
>Lake Kivu: •located in western Rwanda •on the slopes of the Virunga Mountains •in the Great Rift Valley •on the Rwanda-Democratic Republic of the Congo border

POLITICAL:

Independence: July 1, 1962 (from Belgian administration)
Former Name: Ruanda
Bordering Countries: Burundi, Democratic Republic of the Congo, Tanzania, Uganda (4)
Administrative Divisions: Est, Nord, Ouest, Sud (4 provinces) • Kigali (1 city)
Ethnic/Racial Groups: Hutu, Tutsi, Twa
Religions: Christianity (Roman Catholic, Protestant), Islam
Languages: Kinyarwanda, French, English, Kiswahili
Currency: Rwandan franc
Current President: Paul Kagame
Cities (capital, largest, or with at least a million people):
>Kigali: •located in central Rwanda •city in the Kigali-ville prefecture •capital of Rwanda •most populated city in Rwanda (909,000 people) •on the edge of the Virunga Mountains

ENVIRONMENTAL/ECONOMIC:
Climate: temperate with rainy/dry seasons; mild in the mountains with some snow
Natural Resources: gold, tin ore, tungsten ore, methane, hydropower
Agricultural Products: coffee, tea, pyrethrum, bananas, livestock
Major Exports: coffee, tea, hides, tin ore
Natural Hazards: droughts, volcanoes

COUNTRIES

SAINT KITTS AND NEVIS

Country Name: Federation of Saint Kitts and Nevis
Continent: North America
Area: 104 sq mi / 269 sq km (smallest in North America)
Population: 50,000 (smallest in North America)
Population Density: 481 people per sq mi / 186 people per sq km
Capital: Basseterre

PHYSICAL:
Highest Point: Mount Liamuiga 3,793 ft / 1,156 m
Lowest Point: Caribbean Sea 0 ft / 0 m (sea level)
Oceans:
  Atlantic: •forms the eastern coast of Saint Kitts and Nevis
Seas:
  Caribbean: •forms the western coast of Saint Kitts and Nevis •feeds into the Atlantic Ocean
Islands:
  Nevis: •second largest island in Saint Kitts and Nevis •bordered by the Atlantic Ocean and Caribbean Sea

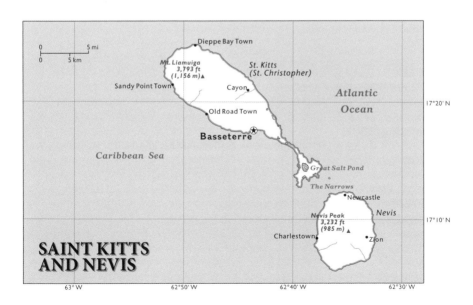

COUNTRIES A TO Z | PAGE 287

Saint Kitts: •largest island in Saint Kitts and Nevis •bordered by the Atlantic Ocean and Caribbean Sea

POLITICAL:
Independence: September 19, 1983 (from the United Kingdom)
Former Name: Federation of Saint Christopher and Nevis
Administrative Divisions: Christ Church Nichola Town, Saint Anne Sandy Point, Saint George Basseterre, Saint George Gingerland, Saint James Windward, Saint John Capesterre, Saint John Figtree, Saint Mary Cayon, Saint Paul Capesterre, Saint Paul Charlestown, Saint Peter Basseterre, Saint Thomas Lowland, Saint Thomas Middle Island, Trinity Palmetto Point (14 parishes)
Ethnic/Racial Groups: black, British, Portuguese, Lebanese
Religion: Christianity (Anglican, other Protestant)
Language: English
Currency: East Caribbean dollar
Current Prime Minister: Denzil Douglas
Cities (capital, largest, or with at least a million people):
Basseterre: •located on the island of Saint Kitts •city in the Saint George Basseterre parish •capital of Saint Kitts and Nevis •most populated city in Saint Kitts and Nevis (13,000 people) •chief port on the Caribbean Sea

ENVIRONMENTAL/ECONOMIC:
Climate: tropical, moderated by sea breezes; little temperature variation; rainy/dry seasons
Natural Resources: arable land
Agricultural Products: sugarcane, rice, yams, vegetables, fish
Major Exports: machinery, food, electronics, beverages, tobacco
Natural Hazards: hurricanes

SAINT LUCIA
Country Name: Saint Lucia
Continent: North America
Area: 238 sq mi / 616 sq km
Population: 176,000
Population Density: 740 people per sq mi / 286 people per sq km
Capital: Castries

PHYSICAL:
Highest Point: Mount Gimie 3,117 ft / 950 m
Lowest Point: Atlantic Ocean 0 ft / 0 m (sea level)
Oceans:
Atlantic: •forms the eastern coast of Saint Lucia
Seas:
Caribbean: •forms the western coast of Saint Lucia •feeds into the Atlantic Ocean

Islands:
   Saint Lucia: •main island in Saint Lucia •bordered by the Atlantic Ocean and Caribbean Sea

POLITICAL:
Independence: February 22, 1979 (from the United Kingdom)
Administrative Divisions: Anse-la-Raye, Castries, Choiseul, Dauphin, Dennery, Gros-Islet, Laborie, Micoud, Praslin, Soufriere, Vieux-Fort (11 quarters)
Ethnic/Racial Groups: black, mixed, East Indian, white
Religion: Christianity (Roman Catholic, Protestant)
Languages: English, French patois
Currency: East Caribbean dollar
Current Prime Minister: Stephenson King
Cities (capital, largest, or with at least a million people):
   Castries: •located in northwestern Saint Lucia •city in the Castries quarter •capital of Saint Lucia •most populated city in Saint Lucia (15,000 people) •chief port on the Caribbean Sea

ENVIRONMENTAL/ECONOMIC:
Climate: tropical, moderated by northeast trade winds; dry/rainy seasons
Natural Resources: forests, sandy beaches, minerals, mineral springs, geothermal potential
Agricultural Products: bananas, coconuts, vegetables, citrus
Major Exports: bananas, clothing, cacao, vegetables
Natural Hazards: hurricanes, volcanoes

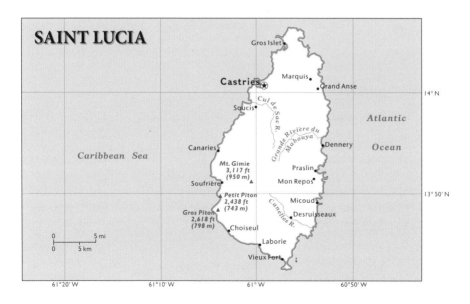

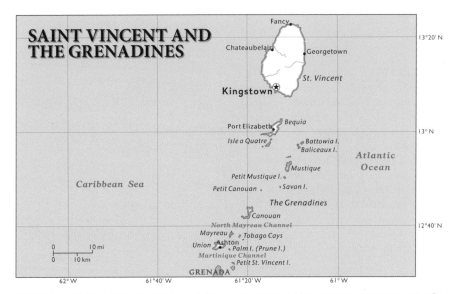

SAINT VINCENT AND THE GRENADINES

Country Name: Saint Vincent and the Grenadines
Continent: North America
Area: 150 sq mi / 389 sq km
Population: 109,000
Population Density: 727 people per sq mi / 280 people per sq km
Capital: Kingstown

PHYSICAL:
Highest Point: Soufriere volcano 4,049 ft / 1,234 m
Lowest Point: Caribbean Sea 0 ft / 0 m (sea level)
Oceans:
   Atlantic: •forms the eastern coast of Saint Vincent and the Grenadines
Seas:
   Caribbean: •forms the western coast of Saint Vincent and the Grenadines
Islands:
   The Grenadines: •island group south of Saint Vincent •bordered by the Atlantic Ocean and Caribbean Sea
   Saint Vincent: •largest island in Saint Vincent and the Grenadines •bordered by the Atlantic Ocean and Caribbean Sea

POLITICAL:
Independence: October 27, 1979 (from the United Kingdom)
Administrative Divisions: Charlotte, Grenadines, Saint Andrew, Saint David, Saint George, Saint Patrick (6 parishes)

Ethnic/Racial Groups: African American, mixed, East Indian, Carib Amerindian
Religion: Christianity (Anglican, Methodist, Roman Catholic, other Protestant)
Languages: English, French patois
Currency: East Caribbean dollar
Current Prime Minister: Ralph E. Gonsalves
Cities (capital, largest, or with at least a million people):
    Kingstown: •located on the island of Saint Vincent •city in the Saint George parish •capital of Saint Vincent and the Grenadines •most populated city in Saint Vincent and the Grenadines (28,000 people) •chief port on the Caribbean Sea

ENVIRONMENTAL/ECONOMIC:
Climate: tropical with little temperature variation; rainy/dry seasons
Natural Resources: hydropower, cropland
Agricultural Products: bananas, coconuts, sweet potatoes, spices, cattle
Major Exports: bananas, eddoes and dasheen (taro), arrowroot starch, tennis racquets
Natural Hazards: hurricanes, volcanoes

SAMOA

Country Name: Independent State of Samoa
Continent: Australia/Oceania
Area: 1,093 sq mi / 2,831 sq km
Population: 191,000
Population Density: 175 people per sq mi / 67 people per sq km
Capital: Apia

PHYSICAL:
Highest Point: Mount Silisili (on the island of Savaii) 6,093 ft / 1,857 m
Lowest Point: Pacific Ocean 0 ft / 0 m (sea level)
Oceans:
    Pacific: •forms most of the coast of Samoa
Islands:
    Savaii: •largest island in Samoa •surrounded by the Pacific Ocean
    Upolu: •second largest island in Samoa •surrounded by the Pacific Ocean

POLITICAL:
Independence: January 1, 1962 (from New Zealand administration)
Former Name: Western Samoa
Administrative Divisions: Aana, Aiga-i-le-Tai, Atua, Faasaleleaga, Gagaemauga, Gagaifomauga, Palauli, Satupaitea, Tuamasaga, Vaa-o-Fonoti, Vaisigano (11 districts)
Ethnic/Racial Groups: Samoan, Euronesian
Religion: predominantly Christian faiths
Languages: Samoan, English
Currency: tala

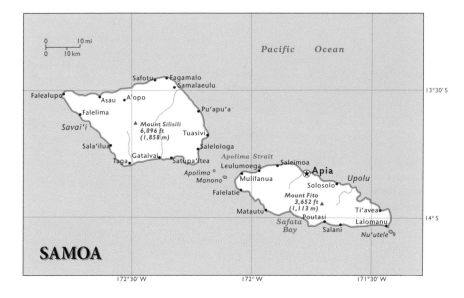

Current Chief: Tuiatua Tupua Tamasese Efi
Cities (capital, largest, or with at least a million people):
  Apia: •located on the island of Upolu •city in the Tuamasaga district •capital of Samoa •most populated city in Samoa (36,000 people) •chief port on the Pacific Ocean

ENVIRONMENTAL/ECONOMIC:
Climate: tropical with rainy/dry seasons
Natural Resources: hardwood forests, fish, hydropower
Agricultural Products: coconuts, bananas, taro, yams
Major Exports: fish, coconut oil and cream, copra, taro
Natural Hazards: typhoons, volcanoes

SAN MARINO
Country Name: Republic of San Marino
Continent: Europe
Area: 24 sq mi / 61 sq km
Population: 32,000
Population Density: 1,333 people per sq mi / 525 people per sq km
Capital: San Marino

PHYSICAL:
Highest Point: Monte Titano 2,425 ft / 739 m
Lowest Point: Torrente Ausa 180 ft / 55 m

Mountain Ranges:
   Apennines: •cover much of the country •extend into Italy

POLITICAL:
Independence: September 3, 301 (established as an independent republic)
Bordering Country: Italy (1)
Administrative Divisions: Acquaviva, Borgo Maggiore, Chiesanuova, Domagnano, Faetano, Fiorentino, Montegiardino, San Marino, Serravalle (9 municipalities)
Ethnic/Racial Groups: Sammarinese, Italian
Religion: Christianity (Roman Catholic)
Language: Italian
Currency: euro
Current Leader: Captains Regent Gabrielle Gatti and Matteo Fiorini
Cities (capital, largest, or with at least a million people):
   Serraville: •located in northeastern San Marino •city in the Serraville municipality •most populated city in San Marino (9,400 people) •on the edge of the Apennines
   San Marino: •located in western San Marino •in the San Marino municipality •capital of San Marino (4,000 people)

ENVIRONMENTAL/ECONOMIC:
Climate: Mediterranean with mild to cool winters/warm and sunny summers
Natural Resources: building stone
Agricultural Products: wheat, grapes, corn, olives, cattle
Major Exports: building stone, lime, wood, chestnuts, wheat
Natural Hazards: landslides

SAO TOME AND PRINCIPE

Country Name: Democratic Republic of Sao Tome and Principe
Continent: Africa
Area: 386 sq mi / 1,001 sq km
Population: 180,000
Population Density: 466 people per sq mi / 180 people per sq km
Capital: Sao Tome

PHYSICAL:
Highest Point: Pico de Sao Tome 6,640 ft / 2,024 m
Lowest Point: Atlantic Ocean 0 ft / 0 m (sea level)
Oceans:
   Atlantic: •forms the southern coast of Sao Tome
Gulfs:
   Gulf of Guinea: •forms all but the southern coast of Sao Tome and Principe •feeds into the Atlantic Ocean
Islands:
   Principe: •second largest island in Sao Tome and Principe •surrounded by the Gulf of Guinea
   Sao Tome: •largest island in Sao Tome and Principe •bordered by the Atlantic Ocean and the Gulf of Guinea

POLITICAL:
Independence: July 12, 1975 (from Portugal)
Administrative Divisions: Principe, Sao Tome (2 provinces)
Ethnic/Racial Groups: mestizo, black, Portuguese

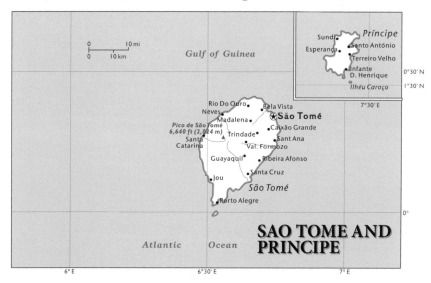

Religion: Christianity (Roman Catholic, Evangelical, Protestant)
Language: Portuguese
Currency: dobra
Current President: Manuel Pinto da Costa
Cities (capital, largest, or with at least a million people):
  Sao Tome: •located on the island of Sao Tome •city in the Sao Tome province •capital of Sao Tome and Principe •most populated city in Sao Tome and Principe (60,000 people) •chief port on the Gulf of Guinea

ENVIRONMENTAL/ECONOMIC:
Climate: tropical hot and humid with rainy/dry seasons
Natural Resources: fish, hydropower
Agricultural Products: cacao, coconuts, palm kernels, copra, poultry
Major Exports: cacao, copra, coffee, palm oil
Natural Hazards: floods

SAUDI ARABIA
Country Name: Kingdom of Saudi Arabia
Continent: Asia
Area: 756,985 sq mi / 1,960,582 sq km
Population: 27,897,000
Population Density: 37 people per sq mi / 14 people per sq km
Capital: Riyadh

PHYSICAL:
Highest Point: Jabal Sawda 10,279 ft / 3,133 m
Lowest Point: Red Sea 0 ft / 0 m (sea level)
Mountain Ranges:
  Jabal al Hijaz: •cover much of western Saudi Arabia •extend into Yemen
Peaks (minimum elevation of 8,000 ft / 2,400 m):
  Jabal Sawda: •located in the Jabal al Hijaz •highest point in Saudi Arabia

Plateau Regions:
  Najd: •covers much of central Saudi Arabia •desert basin west of the Jabal al Hijaz •north of the Rub al Khali
Deserts:
  Rub al Khali: •covers much of southern Saudi Arabia •extends into Yemen, Oman, United Arab Emirates, and Qatar
Seas:
  Red: •forms most of the western coast of Saudi Arabia •feeds into the Gulf of Aden
Gulfs:
  Gulf of Aqaba: •forms part of the northwestern coast of Saudi Arabia •feeds into the Red Sea

Gulf of Bahrain: •forms part of the eastern coast of Saudi Arabia •feeds into the Persian Gulf

Persian: •forms most of the eastern coast of Saudi Arabia •feeds into the Gulf of Oman

POLITICAL:

Independence: September 23, 1932 (date of unification of the kingdom)

Bordering Countries: Yemen, Iraq, Jordan, Oman, United Arab Emirates, Kuwait, Qatar (7)

Regions:

Asir: •region in southwestern Saudi Arabia •includes part of the Jabal al Hijaz •west of the Rub al Khali

Administrative Divisions: Al-Jouf, Asir, Baha, Eastern Region, Hail, Jizan, Madinah, Makkah, Najran, Northern Border, Qasim, Riyadh, Tabouk (13 provinces)

Ethnic/Racial Groups: Arab, Afro-Asian

Religion: predominantly Islam

Language: Arabic

Currency: Saudi riyal

Current King: Abdallah bin Abd al-Aziz Al Saud

Cities (capital, largest, or with at least a million people):

Riyadh: •located in central Saudi Arabia •city in the Ar Riyad province •capital of Saudi Arabia •most populated city in Saudi Arabia (4,725,000 people) •on the eastern edge of the Najd

Jeddah: •located in southwestern Saudi Arabia •city in the Makkah province •chief port on the Red Sea

Mecca: •located in southwestern Saudi Arabia •city in the Makkah province •on the edge of the Jabal al Hijaz

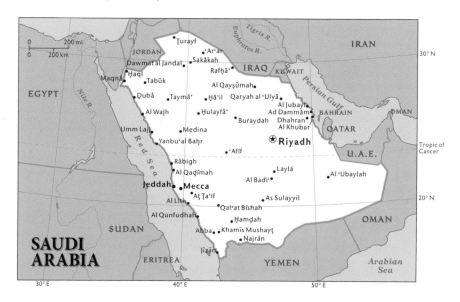

ENVIRONMENTAL/ECONOMIC:
Climate: dry desert with extreme temperatures
Natural Resources: petroleum, natural gas, iron ore, gold, copper
Agricultural Products: wheat, barley, tomatoes, melons, mutton
Major Exports: petroleum and petroleum products
Natural Hazards: sandstorms, dust storms

SENEGAL
Country Name: Republic of Senegal
Continent: Africa
Area: 75,955 sq mi / 196,722 sq km
Population: 12,768,000
Population Density: 168 people per sq mi / 65 people per sq km
Capital: Dakar

PHYSICAL:
Highest Point: near Nepen Diakha 1,906 ft / 581 m
Lowest Point: Atlantic Ocean 0 ft / 0 m (sea level)
Grasslands/Prairies:
  Sahel: •covers most of the country •extends into the Gambia, Mauritania, Mali, Guinea, and Guinea-Bissau
Oceans:
  Atlantic: •forms the western coast of Senegal
Rivers:
  Gambie (Gambia): •extends into Guinea and the Gambia

Senegal: •has its mouth in the Atlantic Ocean •forms all of the Senegal-Mauritania border •extends into Mali

POLITICAL:
Independence: April 4, 1960 (from France)
Bordering Countries: Mauritania, Gambia, Mali, Guinea-Bissau, Guinea (5)
Administrative Divisions: Dakar, Diourbel, Fatick, Kaffrine, Kaolack, Kedougou, Kolda, Louga, Matam, Saint-Louis, Sedhiou,Tambacounda, Thies, Ziguinchor (14 regions)
Ethnic/Racial Groups: Wolof, Pular, Serer, Joia, Mandinka
Religions: Islam, Christianity
Languages: French, Wolof, Pulaar, Diola, Jola, Mandinka
Currency: Communaute Financiere Africaine franc
Current President: Abdoulaye Wade
Cities (capital, largest, or with at least a million people):
Dakar: •located in western Senegal •city in the Dakar region of Senegal •capital of Senegal •most populated city in Senegal (2,777,000 people) •chief port on the Atlantic Ocean

ENVIRONMENTAL/ECONOMIC:
Climate: tropical hot and humid with rainy/dry seasons
Natural Resources: fish, phosphates, iron ore
Agricultural Products: peanuts, millet, corn, sorghum, cattle
Major Exports: fish, groundnuts, petroleum products, phosphates, cotton
Natural Hazards: floods, droughts

SERBIA
Country Name: Serbia
Continent: Europe
Area: 29,913 sq mi / 77,474 sq km
Population: 7,257,000
Population Density: 243 people per sq mi / 94 people per sq km
Capital: Belgrade

PHYSICAL:
Highest Point: Midzor 7,113 ft / 2,168 m
Lowest Point: Danube and Timok Rivers 115 ft / 35 m
Mountain Ranges:
Balkan: • in southeast Serbia •extend into Bulgaria
Grasslands/Prairies:
Great Alfold: •covers most of northern Serbia •extends into Croatia, Hungary, and Romania
Lakes:
Lake Palic

Rivers:
  Sava: • tributary of the Danube River • forms part of the border with Bosnia and Herzegovina
  Danube: • forms Serbia's northeast border
  Drina River: • forms the border between Bosnia and Herzegovina and Serbia

POLITICAL:
Independence: June 5, 2006 (from the federated union of Serbia and Montenegro; was part of Yugoslavia)
Former Name: Serbia and Montenegro, previously Yugoslavia
Bordering Countries: Montenegro, Bosnia and Herzegovina, Croatia, Hungary, Romania, Bulgaria, Macedonia, Kosovo
Administrative Divisions: Serbia is made up of 167 municipalities. For the full list, please visit the CIA World Factbook page at: https://www.cia.gov/library/publications/the-world-factbook/geos/ri.html.
Ethnic/Racial Groups: Serb, Hungarian, Roma, Yugoslavs, Bosniaks, Montenegrin
Religions: Serbian Orthodox, Catholic, Protestant, Muslim
Languages: Serbian, Hungarian, Albanian
Currency: dinar
Current President: Boris Tadic
Cities (capital, largest, or with at least a million people):
  Belgrade: • capital of Serbia (1,115,000 people)

ENVIRONMENTAL/ECONOMIC:
Climate: Continental climate in the north, with cold winters and hot, humid summers and rainfall throughout; in other areas, Continental and Mediterranean climate, with

cold winters, including heavy snowfall, and hot, dry summers and autumns
Natural Resources: oil, gas, coal, iron ore, copper, zinc, antimony, chromite, gold, silver, limestone, marble, salt, arable land
Agricultural: wheat, maize, sugar beets, sunflower, raspberries, beef
Major Exports: manufactured goods, food and live animals, machinery and transportation equipment
Natural Hazards: earthquakes

SEYCHELLES
Country Name: Republic of Seychelles
Continent: Africa
Area: 176 sq mi / 455 sq km (smallest country in Africa)
Population: 88,000 (smallest in Africa)
Population Density: 500 people per sq mi / 193 people per sq km
Capital: Victoria

PHYSICAL:
Highest Point: Morne Seychellois 2,969 ft / 905 m
Lowest Point: Indian Ocean 0 ft / 0 m (sea level)
Oceans:
   Indian: •surrounds the Seychelles
Islands:
   Seychelles: •group of islands extending northeast of Madagascar and Comoros
   Mahe: •largest island in Seychelles •surrounded by the Indian Ocean

POLITICAL:
Independence: June 29, 1976 (from the United Kingdom)
Administrative Divisions: Anse aux Pins, Anse Boileau, Anse Etoile, Anse Royale, Au Cap, Baie Lazare, Baie Sainte Anne, Beau Vallon, Bel Air, Bel Ombre, Cascade, Glacis, Grand' Anse (on Mahe), Grand' Anse (on Praslin), La Digue, La Riviere Anglaise, Les Mamelles, Mont Buxton, Mont Fleuri, Plaisance, Pointe La Rue, Port Glaud, Roche Camain, Saint Louis, Takamaka (25 administrative districts)
Ethnic/Racial Groups: predominantly Seychellois (mix of French, African, and Asian)
Religion: Christianity (Roman Catholic, Anglican)
Languages: English, French, Creole
Currency: Seychelles rupee
Current President: James Alix Michel
Cities (capital, largest, or with at least a million people):
   Victoria: •city on Mahe Island •capital of Seychelles •most populated city in Seychelles (26,000 people) •chief port on the Indian Ocean

ENVIRONMENTAL/ECONOMIC:
Climate: tropical humid marine with warmer/cooler seasons
Natural Resources: fish, copra, cinnamon trees

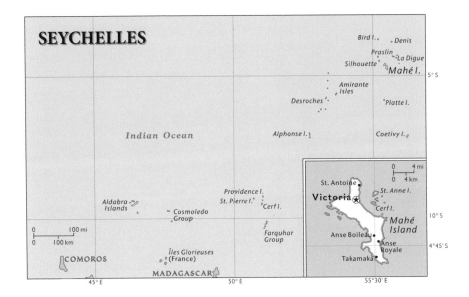

Agricultural Products: coconuts, cinnamon, vanilla, sweet potatoes, broiler chickens
Major Exports: canned tuna, frozen fish, cinnamon bark, copra, petroleum products
Natural Hazards: droughts

SIERRA LEONE
Country Name: Republic of Sierra Leone
Continent: Africa
Area: 27,699 sq mi / 71,740 sq km
Population: 5,364,000
Population Density: 194 people per sq mi / 75 people per sq km
Capital: Freetown

PHYSICAL:
Highest Point: Loma Mansa 6,391 ft / 1,948 m
Lowest Point: Atlantic Ocean 0 ft / 0 m (sea level)
Oceans:
   Atlantic: •forms the western and southern coasts of Sierra Leone

POLITICAL:
Independence: April 27, 1961 (from the United Kingdom)
Bordering Countries: Guinea, Liberia (2)
Administrative Divisions: Eastern, Northern, Southern (3 provinces) •Western (1 area)
Ethnic/Racial Groups: Temne, Mende, other indigenous tribes, Creole
Religions: Islam, indigenous beliefs, Christianity

Languages: English, Mende, Temne, Krio
Currency: leone
Current President: Ernest Bai Koroma
Cities (capital, largest, or with at least a million people):
  Freetown: •located in western Sierra Leone •city in the Western area •capital of Sierra Leone •most populated city in Sierra Leone (875,000 people) •chief port on the Atlantic Ocean

ENVIRONMENTAL/ECONOMIC:
Climate: tropical hot and humid with rainy/dry seasons
Natural Resources: diamonds, titanium ore, bauxite, iron ore, gold
Agricultural Products: rice, coffee, cacao, palm kernels, poultry
Major Exports: diamonds, rutile, cacao, coffee
Natural Hazards: harmattan winds, sandstorms, dust storms

SINGAPORE
Country Name: Republic of Singapore
Continent: Asia
Area: 255 sq mi / 660 sq km
Population: 5,167,000
Population Density: 20,263 people per sq mi / 7,828 people per sq km (most dense in Asia)
Capital: Singapore

PHYSICAL:
Highest Point: Timah 545 ft / 166 m

Lowest Point: South China Sea 0 ft / 0 m (sea level)
Straits:
  Singapore: •forms the southern coast of Singapore •separates Singapore from Indonesia •connects the Strait of Malacca and South China Sea
  Johore: •forms the northern coast of Singapore •separates Singapore from Malaysia •feeds into the Singapore Strait
Islands:
  Singapore: •main island in Singapore •bordered by the Singapore Strait

POLITICAL:

Independence: August 9, 1965 (from the United Kingdom)
Ethnic/Racial Groups: Chinese, Malay, Indian
Religions: Buddhism, Islam, Christianity, Hinduism, Sikh, Taoism, Confucianism
Languages: Chinese, Malay, Tamil, English
Currency: Singapore dollar
Current President: Tony Tan Keng Yam
Cities (capital, largest, or with at least a million people):
  Singapore: •located in southern Singapore •capital of Singapore •most populated city in Singapore (4,737,000 people) •chief port on the Singapore Strait

ENVIRONMENTAL/ECONOMIC:

Climate: tropical hot and humid with monsoon rains
Natural Resources: fish, deep-water ports
Agricultural Products: rubber, copra, fruit, orchids, poultry
Major Exports: machinery and equipment, consumer goods, chemicals, mineral fuels
Natural Hazards: floods, thunderstorms, haze

SLOVAKIA

Country Name: Slovak Republic
Continent: Europe
Area: 18,932 sq mi / 49,035 sq km
Population: 5,440,000
Population Density: 287 people per sq mi / 111 people per sq km
Capital: Bratislava

PHYSICAL:
Highest Point: Gerlachovsky Stit 8,710 ft / 2,655 m
Lowest Point: Bodrok River 308 ft / 94 m
Mountain Ranges:
  Carpathian: •cover much of central Slovakia •extend into Czech Republic, Poland, and Ukraine
Peaks (minimum elevation of 8,000 ft / 2,400 m):
  Gerlachovsky Stit: •located in the Carpathian Mountains •highest point in Slovakia
Grasslands/Prairies:
  Great Alfold: •located in southeastern Slovakia •extends into Ukraine and Hungary
Rivers:
  Danube: •forms part of Slovakia-Hungary border •extends into Austria and Hungary
  Vah: •tributary of the Danube River •has its source in the Carpathian Mountains

PAGE 304 | THE ULTIMATE FACT BOOK

POLITICAL:
Independence: January 1, 1993 (from Czechoslovakia)
Bordering Countries: Hungary, Poland, Czech Republic, Ukraine, Austria (5)
Administrative Divisions: Banskobystricky, Bratislavsky, Kosicky, Nitriansky, Presovsky, Trenciansky, Trnavsky, Zilinsky (8 regions)
Ethnic/Racial Groups: Slovak, Hungarian, Roma
Religion: Christianity (Roman Catholic, Protestant, Orthodox)
Languages: Slovak, Hungarian
Currency: euro
Current President: Ivan Gasparovic
Cities (capital, largest, or with at least a million people):
  Bratislava: •located in southwestern Slovakia •city in the Bratislavsky region •capital of Slovakia •most populated city in Slovakia (428,000 people) •chief port on the Danube River

ENVIRONMENTAL/ECONOMIC:
Climate: temperate with cool summers/cold winters
Natural Resources: brown coal, lignite, iron ore, copper ore, manganese ore
Agricultural Products: grains, potatoes, sugar beets, hops, pigs
Major Exports: machinery and transport equipment, manufactured goods
Natural Hazards: floods, thunderstorms, landslides

SLOVENIA
Country Name: Republic of Slovenia
Continent: Europe
Area: 7,827 sq mi / 20,273 sq km
Population: 2,052,000
Population Density: 262 people per sq mi / 101 people per sq km
Capital: Ljubljana

PHYSICAL:
Highest Point: Triglav 9,396 ft / 2,864 m
Lowest Point: Gulf of Trieste 0 ft / 0 m (sea level)
Mountain Ranges:
  Julian Alps: •cover much of western Slovenia •extend into Italy, Austria, and Croatia
Peaks (minimum elevation of 8,000 ft / 2,400 m):
  Triglav: •located in the Julian Alps •highest point in Slovenia
Gulfs:
  Gulf of Trieste: •forms the southwestern coast of Slovenia •feeds into the Gulf of Venice
Rivers:
  Drava: •extends into Austria, Croatia, and Hungary
  Sava: •has its source in the Julian Alps •extends into Croatia

POLITICAL:
Independence: June 25, 1991 (from Yugoslavia)
Former Names: People's Republic of Slovenia, Socialist Republic of Slovenia
Bordering Countries: Croatia, Austria, Italy, Hungary (4)
Administrative Divisions: Ajdovscina, Beltinci, Benedikt, Bistrica ob Sotli, Bled, Bloke, Bohinj, Borovnica, Bovec, Braslovce, Brda, Brezice, Brezovica, Cankova, Cerklje na Gorenjskem, Cerknica, Cerkno, Cerkvenjak, Crensovci, Crna na Koroskem, Crnomelj, Destrnik, Divaca, Dobje, Dobrepolje, Dobrna, Dobrova-Horjul-Polhov Gradec, Dobrovnik-Dobronak, Dolenjske Toplice, Dol pri Ljubljani, Domzale, Dornava, Dravograd, Duplek, Gorenja Vas-Poljane, Gorisnica, Gornja Radgona, Gornji Grad, Gornji Petrovci, Grad, Grosuplje, Hajdina, Hoce-Slivnica, Hodos-Hodos, Horjul, Hrastnik, Hrpelje-Kozina, Idrija, Ig, Ilirska Bistrica, Ivancna Gorica, Izola-Isola, Jesenice, Jezersko, Jursinci, Kamnik, Kanal, Kidricevo, Kobarid, Kobilje, Kocevje, Komen, Komenda, Kostel, Kozje, Kranjska Gora, Krizevci, Krsko, Kungota, Kuzma, Lasko, Lenart, Lendava-Lendva, Litija, Ljubno, Ljutomer, Logatec, Loska Dolina, Loski Potok, Lovrenc na Pohorju, Luce, Lukovica, Majsperk, Markovci, Medvode, Menges, Metlika, Mezica, Miklavz na Dravskem Polju, Miren-Kostanjevica, Mirna Pec, Mislinja, Moravce, Moravske Toplice, Mozirje, Muta, Naklo, Nazarje, Odranci, Oplotnica, Ormoz, Osilnica, Pesnica, Piran-Pirano, Pivka, Podcetrtek, Podlehnik, Podvelka, Polzela, Postojna, Prebold, Preddvor, Prevalje, Puconci, Race-Fram, Radece, Radenci, Radlje ob Dravi, Radovljica, Ravne na Koroskem, Razkrizje, Ribnica, Ribnica na Pohorju, Rogaska Slatina, Rogasovci, Rogatec, Ruse, Salovci, Selnica ob Dravi, Semic, Sempeter-Vrtojba, Sencur, Sentilj, Sentjernej, Sentjur pri Celju, Sevnica,Sezana, Skocjan, Skofja Loka, Skofljica, Slovenska Bistrica, Slovenske Konjice, Smarje pri Jelsah, Smartno ob Paki, Smartno pri Litiji, Sodrazica, Solcava, Sostanj, Starse, Store, Sveta Ana, Sveti Andraz v Slovenskih Goricah, Sveti Jurij, Tabor, Tisina, Tolmin, Trbovlje, Trebnje, Trnovska Vas, Trzic, Trzin, Turnisce, Velika Polana, Velike Lasce, Verzej, Videm, Vipava, Vitanje, Vodice, Vojnik, Vransko, Vrhnika, Vuzenica, Zagorje ob Savi, Zalec, Zavrc, Zelezniki, Zetale, Ziri, Zirovnica, Zuzemberk, Zrece (182 municipalities)
•Celje, Koper-Capodistria, Kranj, Ljubljana, Maribor, Murska Sobota, Nova Gorica, Nova Mesto, Ptuj, Slovenj Gradec, Velenje (11 urban municipalities)
Ethnic/Racial Groups: Slovene, Croat, Serb, Bosniak
Religion: Christianity (primarily Roman Catholic)
Languages: Slovenian, Serbo-Croatian
Currency: euro
Current President: Danilo Turk
Cities (capital, largest, or with at least a million people):
   Ljubljana: •located in central Slovenia •city in the Ljubljana urban municipality •capital of Slovenia •most populated city in Slovenia (260,000 people) •on the edge of the Julian Alps •chief port on the Sava River

ENVIRONMENTAL/ECONOMIC:
Climate: Mediterranean along the coast; continental in the mountains and valleys with hot summers/cold winters

Natural Resources: lignite coal, lead, zinc, mercury, uranium
Agricultural Products: potatoes, hops, wheat, sugar beets, cattle
Major Exports: manufactured goods, machinery and transport equipment, chemicals, food
Natural Hazards: floods, earthquakes

SOLOMON ISLANDS

Country Name: Solomon Islands
Continent: Australia/Oceania
Area: 10,954 sq mi / 28,370 sq km
Population: 545,000
Population Density: 50 people per sq mi / 19 people per sq km
Capital: Honiara

PHYSICAL:
Highest Point: Mount Pompomanaseu 8,028 ft / 2,447 m
Lowest Point: Pacific Ocean 0 ft / 0 m (sea level)
Peaks (minimum elevation of 8,000 ft / 2,400 m):
　　Mount Pompomanaseu:　●located on the island of Guadalcanal　●highest point in Solomon Islands
Oceans:
　　Pacific:　●forms most of the northern coast of Solomon Islands
Seas:
　　Solomon:　●forms most of the southern coast of Solomon Islands　●feeds into the Coral Sea and Pacific Ocean
　　Coral:　●forms part of the southern coast of Solomon Islands　●feeds into the Pacific Ocean

Islands:
    Guadalcanal: •largest island in Solomon Islands •bordered by the Solomon Sea

POLITICAL:
Independence: July 7, 1978 (from the United Kingdom)
Former Name: British Solomon Islands
Administrative Divisions: Central, Choiseul (Lauru), Guadalcanal, Isabel, Makira, Malaita, Rennell and Bellona, Temotu, Western (9 provinces) •Honiara (1 capital territory)
Ethnic/Racial Groups: Melanesian, Polynesian, Micronesian, European
Religion: Christianity (Protestant, Roman Catholic)
Languages: Melanesian pidgin, indigenous languages, English
Currency: Solomon Islands dollar
Current Prime Minister: Danny Philip
Cities (capital, largest, or with at least a million people):
    Honiara: •located on the island of Guadalcanal •city in the Honiara capital territory •capital of Solomon Islands •most populated city in Solomon Islands (72,000 people) •chief port on the Solomon Sea

ENVIRONMENTAL/ECONOMIC:
Climate: tropical monsoon
Natural Resources: fish, forests, gold, bauxite, phosphates
Agricultural Products: cacao, coconuts, palm kernels, rice, cattle
Major Exports: timber, fish, copra, palm oil
Natural Hazards: typhoons, earthquakes, volcanoes

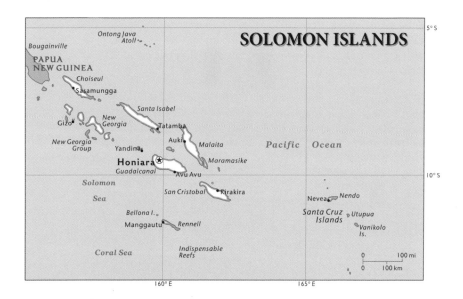

SOMALIA
Country Name: Somalia
Continent: Africa
Area: 246,201 sq mi / 637,657 sq km
Population: 9,926,000
Population Density: 40 people per sq mi / 16 people per sq km
Capital: Mogadishu

PHYSICAL:
Highest Point: Shimbiris 7,927 ft / 2,416 m
Lowest Point: Indian Ocean 0 ft / 0 m (sea level)
Oceans:
  Indian: •forms the eastern coast of Somalia
Gulfs:
  Gulf of Aden: •forms the northern coast of Somalia •feeds into the Arabian Sea and Indian Ocean
Rivers:
  Webi Shabeelle: •has its mouth in the Indian Ocean •extends into Ethiopia
Regions:
  Haud: •lowland, desert region that stretches across northern Somalia •extends into Ethiopia

POLITICAL:
Independence: July 1, 1960 (from United Kingdom and Italian administration)
Former Names: Somali Republic, Somali Democratic Republic, British Somaliland, Italian Somaliland
Bordering Countries: Ethiopia, Kenya, Djibouti (3)
Regions:
  Somaliland •mountainous region in northern Somalia •declared independence from Somalia in 1991
Administrative Divisions: Awdal, Bakool, Banaadir, Bari, Bay, Galguduud, Gedo, Hiiraan, Jubbada Dhexe, Jubbada Hoose, Mudug, Nugaal, Sanaag, Shabeellaha Dhexe, Shabeellaha Hoose, Sool, Togdheer, Woqooyi Galbeed (18 regions)
Ethnic/Racial Groups: Somali, Bantu
Religion: Islam (primarily Sunni)
Languages: Somali, Arabic, Italian, English
Currency: Somali shilling
Current President: Sheikh Sharif Sheikh Ahmed
Cities (capital, largest, or with at least a million people):
  Mogadishu: •located in southeastern Somalia •city in the Banaadir region •capital of Somalia •most populated city in Somalia (1,353,000 people) •chief port on the Indian Ocean

ENVIRONMENTAL/ECONOMIC:
Climate: mainly desert with mild to hot temperatures and two monsoon seasons
Natural Resources: uranium, iron ore, tin, gypsum, bauxite
Agricultural Products: bananas, sorghum, corn, coconuts, cattle
Major Exports: livestock, bananas, hides, fish, charcoal
Natural Hazards: droughts, dust storms, floods

SOUTH AFRICA
Country Name: Republic of South Africa
Continent: Africa
Area: 470,693 sq mi / 1,219,090 sq km
Population: 50,460,000
Population Density: 107 people per sq mi / 41 people per sq km
Capitals: Pretoria (administrative capital), Cape Town (legislative capital), Bloemfontein (judicial capital)

PHYSICAL:
Highest Point: Mafadi 11,306 ft / 3,446 m
Lowest Point: Atlantic Ocean 0 ft / 0 m
Mountain Ranges:
  Drakensberg: •cover an area from southern to northeastern South Africa
Peaks (minimum elevation of 8,000 ft / 2,400 m):
  Mafadi: •located in the Drakensberg •on the South Africa–Lesotho border (just

north of Giant's Castle) •highest point in South Africa
Plateau Regions:
  Great Karoo: •located in southern South Africa •basin at the western end of the Drakensberg
Deserts:
  Kalahari: •located in northern and western South Africa •extends into Namibia and Botswana
Oceans:
  Atlantic: •forms part of the western coast of South Africa
  Indian: •forms the eastern coast and most of the southern coast of South Africa
Rivers:
  Limpopo: •has its source in the Drakensberg •forms all of the South Africa–Zimbabwe border •forms part of the South Africa–Botswana border •extends into Mozambique
  Orange: •has its mouth in the Atlantic Ocean •forms much of the South Africa-Namibia border •extends into Lesotho
  Vaal: •tributary of the Orange River •has its source in the Drakensberg
Capes:
  Cape Agulhas: •bordered by the Indian Ocean •southernmost point in Africa
  Cape of Good Hope: •bordered by the Atlantic and Indian Oceans

POLITICAL:
Independence: May 31, 1910 (from the United Kingdom)
Former Name: Union of South Africa
Bordering Countries: Botswana, Namibia, Lesotho, Mozambique, Swaziland, Zimbabwe (6)

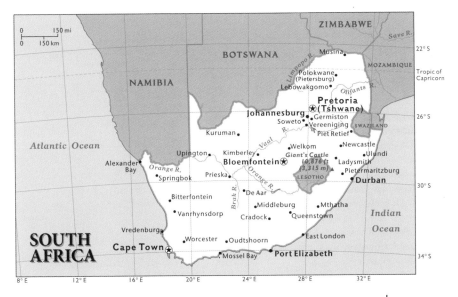

Regions:
    Bushmanland: •region in western South Africa •east of the Atlantic Ocean
    Zululand: •region in eastern South Africa •west of the Indian Ocean
Administrative Divisions: Eastern Cape, Free State, Gauteng, KwaZulu-Natal, Limpopo, Mpumalanga, Northern Cape, North-West, Western Cape (9 provinces)
Ethnic/Racial Groups: black, white, mixed, Indian
Religions: Christianity, indigenous beliefs, Islam, Hinduism
Languages: Afrikaans, English, Ndebele, Pedi, Sotho, Swazi, Tsonga, Tswana, Venda, Xhosa, Zulu
Currency: rand
Current President: Jacob Zuma
Cities (capital, largest, or with at least a million people):
    Johannesburg: •located in northeastern South Africa •city in the Gauteng province •most populated city in South Africa (3,670,000 people)
    Cape Town: •located in southwestern South Africa •city in the Western Cape province •legislative capital of South Africa (3,353,000 people)•a major port on the Atlantic Ocean
    Durban: •located in eastern South Africa •city in the KwaZulu-Natal province •on the edge of the Drakensberg •a major port on the Indian Ocean
    Pretoria: •located in northeastern South Africa •city in the Gauteng province •administrative capital of South Africa (1,404,000 people)
    Soweto: •located in northeastern South Africa •city in the Gauteng province
    Bloemfontein: •located in central South Africa •city in the Free State province •judicial capital of South Africa (436,000 people)

ENVIRONMENTAL/ECONOMIC:
Climate: mostly semi-arid; subtropical on the east coast
Natural Resources: gold, chromium, antimony, coal, iron ore
Agricultural Products: corn, wheat, sugarcane, fruits, beef
Major Exports: gold, diamonds, platinum, other metals and minerals, machinery and equipment
Natural Hazards: droughts

SOUTH SUDAN
Country Name: South Sudan
Continent: Africa
Area: 248,777 sq mi / 644,329 sq km
Population: 8,260,000
Population Density: 33 people per sq mi / 13 people per sq km
Capital: Juba

PHYSICAL:
Highest Point: Kinyeti 100,456 ft / 3,187 m
Lowest Point: NA

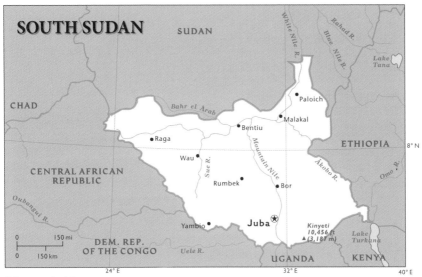

Mountain Ranges:
  Imatong: • located in the southeastern part of South Sudan • extend into Uganda
  Dongotona: • located in the southeastern part of South Sudan • extend into Uganda
Peaks (minimum elevation of 8,000 ft/2,400 m):
  Kinyeti: • highest point in South Sudan
Grasslands/Prairies:
  Sahel: • located in the northern part of South Sudan • extends from Senegal in eastern Africa to Chad, Central African Republic, Sudan, and South Sudan
Rivers:
  White Nile: • tributary of the Nile River • flows north out of Central Africa • extends into Uganda
Swamps:
  The Sudd: • Saharan flooded grasslands • one of the largest swamps in the world

POLITICAL:
Independence: July 9, 2011
Former Name: Sudan
Bordering Countries: Sudan, Ethiopia, Kenya, Uganda, Democratic Republic of Congo, Central African Republic
Administrative Divisions: Central Equatoria, Eastern Equatoria, Jonglei, Lakes, Northern Bahr el Ghazal, Unity, Upper Nile, Warrap, Western Bahr el Ghazal, Western Equatoria (10 states)
Ethnic/Racial Groups: Dinka, Kakwa, Bari, Azande, Shilluk, Kuku, Murle, Mandari, Didinga, Ndogo, Bviri, Lndi, Anuak, Bongo, Lango, Dungotona, Acholi
Religions: animist, Christian
Languages: English, Arabic, regional languages (Dinka, Nuer, Bari, Zande, Shilluk)

Currency: South Sudan pound
Current President: Salva Kiir Mayardit
Cities (capital, largest, or with at least a million people):
  Juba: •located in the southern part of the country •capital of South Sudan •largest city in South Sudan (250,000 people)

ENVIRONMENTAL/ECONOMIC:
Climate: Equatorial climate, hot with seasonal rainfall, which is heaviest in the highlands areas of the south
Natural Resources: hydropower, fertile agricultural land, gold, diamonds, petroleum, hardwoods, limestone, iron ore, copper, chromium ore, zinc, tungsten, mica, silver
Agricultural: sorghum, maize, rice, millet, wheat, gum Arabic, sugarcane, mangoes, papayas, bananas, sweet potatoes, sunflowers, cotton, sesame, cassava, beans, peanuts, cattle, sheep

SPAIN
Country Name: Kingdom of Spain
Continent: Europe
Area: 195,363 sq mi / 505,988 sq km
Population: 47,262,000
Population Density: 242 people per sq mi / 93 people per sq km
Capital: Madrid

PHYSICAL:
Highest Point: Pico de Teide (on Tenerife) 12,198 ft / 3,718 m
Lowest Point: Atlantic Ocean 0 ft / 0 m (sea level)
Mountain Ranges:
    Cordillera Cantabrica (Cantabrian): • cover much of northern Spain
    Pyrenees: • located in northeastern Spain • extend into France and Andorra
    Sierra Morena: • cover much of southern Spain
Peaks (minimum elevation of 8,000 ft / 2,400 m):
    Pico de Teide: • located on the island of Tenerife in the Canary Islands • highest point in Spain
Oceans:
    Atlantic: • forms part of the western coast of Spain
Seas:
    Alboran (sometimes considered part of the Mediterranean): • forms much of the southern coast of Spain • feeds into the Atlantic Ocean
    Balearic: • forms much of the eastern coast of Spain • feeds into the Mediterranean Sea
    Mediterranean: • forms much of the eastern and southern coasts of Spain • feeds into the Atlantic Ocean
Gulfs:
    Gulf of Cadiz: • forms part of the southwestern coast of Spain • feeds into the Atlantic Ocean
    Gulf of Valencia: • forms part of the eastern coast of Spain • feeds into the Balearic Sea
Bays:
    Bay of Biscay: • forms most of the northern coast of Spain • feeds into the Atlantic Ocean
Straits:
    Strait of Gibraltar: • forms part of the southwestern coast of Spain • separates Spain from Morocco • connects the Alboran Sea and the Atlantic Ocean
Rivers:
    Duero (Douro): • has its source in northern Spain • forms part of the Spain-Portugal border • extends into Portugal
    Ebro: • has its mouth in the Balearic Sea • has its source in the Cordillera Cantabrica
    Guadiana: • has its mouth in the Gulf of Cadiz • has its source in central Spain • forms part of the Spain-Portugal border • extends into Portugal
    Tagus: • has its source in eastern Spain • forms part of the Spain-Portugal border • extends into Portugal
Islands:
    Balearic: • island group between the Mediterranean and Balearic Seas
    Canary: • island group off the the coast of Western Sahara (claimed by Morocco) • surrounded by the Atlantic Ocean
    Gran Canaria: • island in the Canary Islands • surrounded by the Atlantic Ocean
    Majorca: • largest island in Spain • island in the Balearic Islands • bordered by the Mediterranean and Balearic Seas
    Tenerife: • second largest island in Spain • island in the Canary Islands • surrounded

by the Atlantic Ocean

POLITICAL:
Independence: 1492 (final victory over the Moors)
Bordering Countries: Portugal, France, Andorra, Morocco, Gibraltar (dependency of the United Kingdom) (5)
Administrative Divisions: Andalusia, Aragon, Asturias, Baleares (Balearic Islands), Canarias (Canary Islands), Cantabria, Castile-La Mancha, Castile and Leon, Cataluna, Ceuta (along north coast of Morocco), Extremadura, Galicia, La Rioja, Madrid, Melilla (along north coast of Morocco), Murcia, Navarra, Pais Vasco (Basque Country), Valencia (19 autonomous communities)
Ethnic/Racial Groups: Castilian, Catalan, Basque, Galician
Religion: Christianity (primarily Roman Catholic)
Languages: Castilian Spanish, Catalan, Galician, Basque
Currency: euro
Current Prime Minister: Jose Luis Rodriguez Zapatero
Cities (capital, largest, or with at least a million people):
  Madrid: •located in central Spain •city in the Madrid autonomous community •capital of Spain •most populated city in Spain (5,762,000 people)
  Barcelona: •located in northeastern Spain •city in the Cataluna autonomous community •chief port on the Mediterranean Sea
  Valencia: •located in eastern Spain •city in the Communidad Valencian autonomous community •chief port on the Gulf of Valencia
  Seville: •located in southern Spain •city in the Andalucia autonomous community

ENVIRONMENTAL/ECONOMIC:
Climate: temperate with hot summers/cold winters in the interior and moderate summers/cool winters along the coast
Natural Resources: coal, lignite, iron ore, copper, lead
Agricultural Products: grain, vegetables, olives, wine grapes, beef
Major Exports: machinery, motor vehicles, foodstuffs, other consumer goods
Natural Hazards: droughts

SRI LANKA
Country Name: Democratic Socialist Republic of Sri Lanka
Continent: Asia
Area: 25,299 sq mi / 65,525 sq km
Population: 20,858,000
Population Density: 824 people per sq mi / 318 people per sq km
Capital: Colombo

PHYSICAL:
Highest Point: Pidurutalagala 8,281 ft / 2,524 m
Lowest Point: Indian Ocean 0 ft / 0 m (sea level)

Peaks (minimum elevation of 8,000 ft / 2,400 m):
Pidurutalagala: •located in southern Sri Lanka •highest point in Sri Lanka
Oceans:
Indian: •forms the southern coast of Sri Lanka
Gulfs:
Gulf of Mannar: •forms much of the western coast of Sri Lanka •feeds into the Indian Ocean
Bays:
Bay of Bengal: •forms much of the eastern coast of Sri Lanka •feeds into the Indian Ocean
Straits:
Palk: •forms much of the northwestern coast of Sri Lanka •separates the island of Sri Lanka from India •connects the Gulf of Mannar and the Bay of Bengal
Islands:
Sri Lanka: •main island in Sri Lanka •bordered by the Indian Ocean, Gulf of Mannar, Bay of Bengal, and Palk Strait

POLITICAL:
Independence: February 4, 1948 (from the United Kingdom)
Former Names: Serendib, Ceylon
Administrative Divisions: Central, Eastern, North Central, North Eastern, North Western, Sabaragamuwa, Southern, Uva, Western (9 provinces)
Ethnic/Racial Groups: Sinhalese, Tamil, Moor
Religions: Buddhism, Hinduism, Christianity, Islam
Languages: Sinhala, Tamil, English

Currency: Sri Lankan rupee
Current President: Dissanayake Mudiyanselage Jayaratne
Cities (capital, largest, or with at least a million people):
  Colombo: •located in southwestern Sri Lanka •city in the Western province
    •capital of Sri Lanka •most populated city in Sri Lanka (681,000 people)
    •chief port on the Indian Ocean
  Kotte: •administrative capital of Sri Lanka (123,000 people)

ENVIRONMENTAL/ECONOMIC:
Climate: tropical monsoon with two monsoon seasons
Natural Resources: limestone, graphite, mineral sands, gems, phosphates
Agricultural Products: rice, sugarcane, grains, pulses, milk
Major Exports: textiles and apparel, tea, diamonds, coconut products, petroleum products
Natural Hazards: cyclones, tornadoes

SUDAN
Country Name: Republic of the Sudan
Continent: Africa
Area: 718,700 sq mi / 1,861,484 sq km
Population: 44,632,000
Population Density: 62 people per sq mi / 24 people per sq km
Capital: Khartoum

PHYSICAL:
Highest Point: Jebel Marra 10,075 ft / 3,071 m

Lowest Point: Red Sea 0 ft / 0 m (sea level)
Peaks (minimum elevation of 8,000 ft / 2,400 m):
  Jebel Marra: •located in western Sudan •highest point in Sudan
Grasslands/Prairies:
  Sahel: •covers much of central Sudan •extends into South Sudan, Chad, and Central African Republic
Deserts:
  Libyan: •located in northwestern Sudan •extends into Egypt, Libya, and Chad
  Nubian: •covers much of northern Sudan •extends into Egypt
  Sahara: •covers most of northern Sudan •extends into Chad, Libya, and Egypt •largest desert in the world

Seas:
  Red: •forms the eastern coast of Sudan •feeds into the Gulf of Aden
Lakes:
  Lake Nasser: •also known in Sudan as Lake Nubia •located in northern Sudan •fed by the Nile River •formed by the Aswan High Dam •on the edge of the Nubian Desert •surrounded by the Sahara •extends into Egypt
Rivers:
  Blue Nile: •tributary of the Nile River •extends into Ethiopia
  Nile: •extends into Egypt •longest river in the world (including its White Nile tributary)
  White Nile: •tributary of the Nile River •extends into South Sudan and Uganda

POLITICAL:
Independence: January 1, 1956 (from Egypt and the United Kingdom)
Former Name: Anglo-Egyptian Sudan
Bordering Countries: Ethiopia, South Sudan, Central African Republic, Chad, Libya, Egypt, Eritrea (7)
Administrative Divisions: Al Bahr al Ahmar, Al Jazirah, Al Khartoum, Al Qadarif, An Nil al Abyad, An Nil al Azraq, Ash Shimaliyya, Gharb Darfur, Janub Darfur, Janub Kurdufan, Kassala, Nahr an Nil, Shimal Darfur, Shimal Kurdufan, Sinnar (15 states)
Ethnic/Racial Groups: Arab, Beja, Fur, Nuba, Fullata
Religions: Islam (primarily Sunni), indigenous beliefs, Christianity
Languages: Arabic, English, Nubian, Ta Bedawie, many local dialects
Currency: Sudanese pound
Current President: Umar Hassan al-Bashir
Cities (capital, largest, or with at least a million people):
  Khartoum: •located in eastern Sudan •city in the Al Khartum state •capital of Sudan •most populated city in Sudan (5,021,000 people) •chief port on the Nile, White Nile, and Blue Nile Rivers

ENVIRONMENTAL/ECONOMIC:
Climate: arid desert; rainy/dry seasons
Natural Resources: petroleum, iron ore, copper, chromium ore, zinc
Agricultural Products: cotton, groundnuts, sorghum, millet, wheat, silver, mica, gold, hydropower
Major Exports: oil and petroleum products, cotton, sesame, livestock, groundnuts, sugar, gum arabic
Natural Hazards: dust storms, droughts

SURINAME
Country Name: Republic of Suriname
Continent: South America
Area: 63,037 sq mi / 163,265 sq km (smallest in South America)
Population: 529,000 (smallest in South America)
Population Density: 8 people per sq mi / 3 people per sq km (least dense in South America)
Capital: Paramaribo

PHYSICAL:
Highest Point: Juliana Top 4,035 ft / 1,230 m
Lowest Point: unnamed location 7 ft / 2 m below sea level
Mountain Ranges:
   Guiana Highlands: •cover much of southern Suriname •extend into Guyana, Brazil, and French Guiana (overseas territory of France)

Oceans:
  Atlantic: • forms the northern coast of Suriname
Lakes:
  Van Blommestein Meer: • located in eastern Suriname • reservoir of the Suriname River • formed by the Afobaka Dam
  Wonotobo Vallen: • located in western Suriname • on the edge of the Guiana Highlands

POLITICAL:
Independence: November 25, 1975 (from the Netherlands)
Former Names: Netherlands Guiana, Dutch Guiana
Bordering Countries: Guyana, Brazil, French Guiana (French overseas territory) (3)
Administrative Divisions: Brokopondo, Commewijne, Coronie, Marowijne, Nickerie, Para, Paramaribo, Saramacca, Sipaliwini, Wanica (10 districts)
Ethnic/Racial Groups: East Indian, Creole, Javanese, Maroons, Amerindian, Chinese
Religions: Hinduism, Christianity (Protestant, Roman Catholic), Islam
Languages: Dutch, English, Sranang Tongo, Hindustani, Javanese
Currency: Suriname dollar
Current President: Desire Delano Bouterse
Cities (capital, largest, or with at least a million people):
  Paramaribo: • located in northern Suriname • city in the Paramaribo district • capital of Suriname • most populated city in Suriname (259,000 people) • chief port on the Atlantic Ocean

ENVIRONMENTAL/ECONOMIC:
Climate: tropical, moderated by trade winds
Natural Resources: timber, hydropower, fish, kaolin, shrimp
Agricultural Products: paddy rice, bananas, palm kernels, beef, forest products

Major Exports: alumina, crude oil, lumber, shrimp and fish, rice
Natural Disasters: floods, forest fires

SWAZILAND
Country Name: Kingdom of Swaziland
Continent: Africa
Area: 6,704 sq mi / 17,363 sq km
Population: 1,203,000
Population Density: 179 people per sq mi / 69 people per sq km
Capitals: Mbabane (administrative capital), Lobamba (legislative and royal capital)

PHYSICAL:
Highest Point: Mlembe 6,109 ft / 1,862 m
Lowest Point: Great Usutu River 69 ft / 21 m

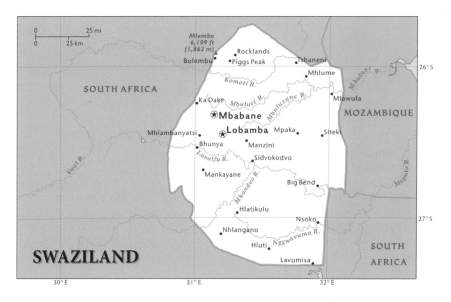

Mountain Ranges:
  Drakensberg: •cover much of western Swaziland •extend into South Africa

POLITICAL:
Independence: September 6, 1968 (from the United Kingdom)
Bordering Countries: South Africa, Mozambique (2)
Administrative Divisions: Hhohho, Lubombo, Manzini, Shiselweni (4 districts)
Ethnic/Racial Groups: native African, European
Religions: indigenous beliefs, Christianity (Roman Catholic), Islam
Languages: English, siSwati
Currency: lilangeni
Current King: Mswati III
Cities (capital, largest, or with at least a million people):
  Mbabane: •located in northwestern Swaziland •in the Hhohho district •judicial capital of Swaziland (74,000 people)
  Lobamba: •in the Hhohho district • legislative and royal capital of Swaziland (4,400 people)

ENVIRONMENTAL/ECONOMIC:
Climate: tropical to temperate
Natural Resources: asbestos, coal, clay, cassiterite, hydropower
Agricultural Products: sugarcane, cotton, corn, tobacco, cattle
Major Exports: soft drink concentrates, sugar, wood pulp, cotton yarn
Natural Hazards: droughts

SWEDEN

Country Name: Kingdom of Sweden
Continent: Europe
Area: 173,732 sq mi / 449,964 sq km
Population: 9,447,000
Population Density: 54 people per sq mi / 21 people per sq km
Capital: Stockholm

PHYSICAL:
Highest Point: Kebnekaise 6,926 ft / 2,111 m
Lowest Point: Lake Hammarsjon 8 ft / 2.4 m below sea level
Mountain Ranges:
  Kjolen: •cover an area from northern to southwestern Sweden •extend into Norway and Finland
Seas:
  Baltic: •forms much of the southern coast of Sweden •feeds into the Kattegat
Gulfs:
  Gulf of Bothnia: •forms part of the eastern coast of Sweden •feeds into the Baltic Sea
  Kattegat: •forms part of the southwestern coast of Sweden •feeds into the North Sea
Straits:
  Kalmar: •forms much of the southeastern coast of Sweden •separates the island of Oland from Sweden
  Oresund: •forms part of the southwestern coast of Sweden •separates Sweden from Denmark •connects the Baltic Sea and the Kattegat

Channels:
    Skagerrak: • forms part of the southwestern coast of Sweden • separates Sweden and Norway from Denmark • connects the Kattegat and the North Sea

Lakes:
    Vanern: • located in southwestern Sweden • largest lake in Sweden
    Vattern: • located in southern Sweden

Islands:
    Gotland: • island in the Baltic Sea • off the southeastern coast of Sweden
    Oland: • island off the coast of southeastern Sweden • bordered by the Baltic Sea and the Kalmar Strait

POLITICAL:

Independence: June 6, 1523 (date Gustav Vasa became king)
Bordering Countries: Norway, Finland (2)

Regions:
    Gotaland: • region that covers much of southern Sweden • east of the Skagerrak, Kattegat, and Oresund • west of the Baltic Sea and Kalmar Strait • includes parts of Lakes Vanern and Vattern
    Lapland: • region located in northern Sweden • includes part of the Kjolen • extends into Norway and Finland
    Norrland: • region that covers much of northern and central Sweden • includes part of the Kjolen • west of the Gulf of Bothnia
    Svealand: • region that covers much of central Sweden • includes part of the Kjolen • west of the Gulf of Bothnia and Baltic Sea • includes parts of Lakes Vanern and Vattern

Administrative Divisions: Blekinge, Dalarnas, Gavleborgs, Gotlands, Hallands, Jamtlands, Jonkopings, Kalmar, Kronobergs, Norrbottens, Orebro, Ostergotlands, Skane, Sodermanlands, Stockholms, Uppsala, Varmlands, Vasterbottens, Vasternorrlands, Vastmanlands, Vastra Gotalands (21 counties)
Ethnic/Racial Groups: Swedish, Finnish, Sami
Religion: Christianity (Lutheran, Roman Catholic)
Language: primarily Swedish
Currency: Swedish krona
Current Prime Minister: Fredrik Reinfeldt
Cities (capital, largest, or with at least a million people):
    Stockholm: • located in southeastern Sweden • city in Stockholms county • capital of Sweden • most populated city in Sweden (1,279,000 people) • chief port on the Baltic Sea

ENVIRONMENTAL/ECONOMIC:

Climate: temperate in the south; sub-arctic in the north; cold winters/cool summers
Natural Resources: iron ore, copper, lead, zinc, gold
Agricultural Products: barley, wheat, sugar beets, meat

Major Exports: machinery, motor vehicles, paper products, pulp and wood, iron and steel products
Natural Hazards: sea ice

SWITZERLAND

Country Name: Swiss Confederation
Continent: Europe
Area: 15,940 sq mi / 41,284 sq km
Population: 7,868,000
Population Density: 494 people per sq mi / 191 people per sq km
Capital: Bern

PHYSICAL:
Highest Point: Dufourspitze 15,203 ft / 4,634 m
Lowest Point: Lake Maggiore 640 ft / 195 m
Mountain Ranges:
  Alps: ●cover much of the country ●extend into France, Italy, Germany, Liechtenstein, and Austria
  Jura: ●cover an area from southwestern to northern Switzerland ●extend into France
Peaks (minimum elevation of 8,000 ft / 2,400 m):
  Dom: ●located in the Alps
  Dufourspitze: ●located in the Alps ●on the Switzerland-Italy border just east of the Matterhorn ●highest point in Switzerland
  Matterhorn: ●located in the Alps ●on the Switzerland-Italy border

COUNTRIES A TO Z | PAGE 325

Passes:
    Great Saint Bernard: •located in the Alps •crosses the border into Italy
Lakes:
    Lake Constance: •located in northeastern Switzerland •on the border with Germany and Austria •on the edge of the Alps •fed by the Rhine River
    Lake of Geneva: •located in southwestern Switzerland •on the Switzerland-France border •between the Jura Mountains and the Alps •fed by the Rhone River
    Lake of Lucerne: •located in central Switzerland •on the edge of the Alps
    Lake Maggiore: •located in southern Switzerland •on the Switzerland-Italy border •on the edge of the Alps
    Lake Neuchatel: •located in western Switzerland •on the edge of the Jura Mountains
Rivers:
    Inn: •has its source in the Alps •extends into Austria
    Rhine: •has its source in the Alps •forms much of Switzerland's border with Germany and Liechtenstein •forms part of the Switzerland-Austria border •extends into Austria and Germany
    Rhone: •has its source in the Alps •extends into France

POLITICAL:
Independence: August 1, 1291 (founding of the Swiss Confederation)
Bordering Countries: Italy, France, Germany, Austria, Liechtenstein (5)
Administrative Divisions: Aargau, Appenzell Inner-Rhoden, Ausserrhoden, Basel-Landschaft, Basel-Stadt, Bern, Fribourg, Geneve, Glarus, Graubunden, Jura, Luzern, Neuchatel, Nidwalden, Obwalden, Sankt Gallen, Schaffhausen, Schwyz, Solothurn, Thurgau, Ticino, Uri, Valais, Vaud, Zug, Zurich (26 cantons)
Ethnic/Racial Groups: German, French, Italian, Romanche
Religion: Christianity (Roman Catholic, Protestant)
Languages: German, French, Italian, Romanisch
Currency: Swiss franc
Current President: Micheline Calmy-Rey
Cities (capital, largest, or with at least a million people):
    Zurich: •located in northern Switzerland •city in the Zurich canton •most populated city in Switzerland (1,150,000 people)
    Bern: •located in western Switzerland •capital of Switzerland (346,000 people) •in the Bern canton

ENVIRONMENTAL/ECONOMIC:
Climate: temperate that varies with altitude; cold and snowy winters/cool to warm summers
Natural Resources: hydropower potential, timber, salt
Agricultural Products: grains, fruits and vegetables, meat
Major Exports: machinery, chemicals, metals, watches, agricultural products
Natural Hazards: avalanches, landslides, flash floods

SYRIA

Country Name: Syrian Arab Republic
Continent: Asia
Area: 71,498 sq mi / 185,180 sq km
Population: 22,518,000
Population Density: 315 people per sq mi / 122 people per sq km
Capital: Damascus

PHYSICAL:

Highest Point: Mount Hermon 9,232 ft / 2,814 m
Lowest Point: NA
Mountain Ranges:
  Anti-Lebanon: •located in southwestern Syria •extend into Lebanon
Peaks (minimum elevation of 8,000 ft / 2,400 m):
  Mount Hermon: •located in the Anti-Lebanon Mountains •on the Syria-Lebanon border •highest point in Syria
Deserts:
  Syrian: •covers much of the country •extends into Jordan and Iraq
Seas:
  Mediterranean: •forms the western coast of Syria •feeds into the Atlantic Ocean
Rivers:
  Euphrates: •extends into Turkey and Iraq
  Orontes: •has its source in the Anti-Lebanon Mountains •extends into Lebanon and Turkey

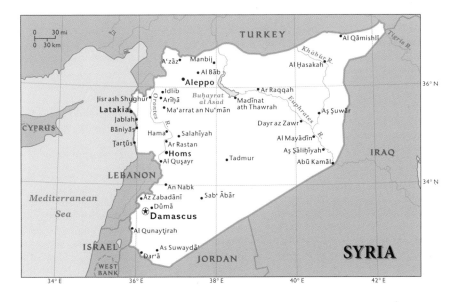

POLITICAL:

Independence: April 17, 1946 (from French administration)
Former Name: United Arab Republic (with Egypt)
Bordering Countries: Turkey, Iraq, Jordan, Lebanon, Israel (5)
Regions:
   Mesopotamia: •ancient region that covers much of eastern Syria •extends into Iraq
Administrative Divisions: Al Hasakah, Al Ladhiqiyah, Al Qunaytirah, Ar Raqqah, As Suwayda, Dara, Dayr az Zawr, Dimashq, Halab, Hamah, Hims, Idlib, Rif Dimashq, Tartus (14 provinces)
Ethnic/Racial/Racial Groups: Arabs, Kurds, Armenians
Religions: Islam (Sunni, Alawite, Druze, other Muslim sects), Christianity
Languages: Arabic, Kurdish, Armenian, Aramaic, Circassian, French, English
Currency: Syrian pound
Current President: Bashar al-Asad
Cities (capital, largest, or with at least a million people):
   Aleppo: •located in northwestern Syria •city in the Halab province
       •most populous city in Syria (3,087,000 people)
   Damascus: •located in southwestern Syria •city in the Dimashq province
       •capital of Syria (2,527,000 people) •between the Anti-Lebanon Mountains and the Syrian Desert

ENVIRONMENTAL/ECONOMIC:

Climate: mostly desert with hot and dry summers/mild and rainy winters along the coast; occasional snow or sleet around Damascus
Natural Resources: petroleum, phosphates, chrome ore, manganese ore, asphalt
Agricultural Products: wheat, barley, cotton, lentils, beef
Major Exports: crude oil, petroleum products, fruits and vegetables, cotton fiber, clothing
Natural Hazards: dust storms, sandstorms

TAJIKISTAN

Country Name: Republic of Tajikistan
Continent: Asia
Area: 55,251 sq mi / 143,100 sq km
Population: 7,535,000
Population Density: 136 people per sq mi / 53 people per sq km
Capital: Dushanbe

PHYSICAL:

Highest Point: Qullai Ismoili Somoni 24,590 ft / 7,495 m
Lowest Point: Syr Darya (river) 984 ft / 300 m
Mountain Ranges:
  Pamirs: •located in eastern Tajikistan •extend into Afghanistan, China, and Kyrgyzstan
Peaks (minimum elevation of 8,000 ft / 2,400 m):
  Qullai Ismoili Somoni •located in the Pamirs •highest point in Tajikistan
Rivers:
  Amu Darya: •has its source in the Pamirs •forms part of the Tajikistan-Afghanistan border •extends into Afghanistan and Uzbekistan
  Syr Darya: •extends into Uzbekistan and Kyrgyzstan •forms part of the Tajikistan-Uzbekistan border

POLITICAL:

Independence: September 9, 1991 (from the Soviet Union)
Former Name: Tajik Soviet Socialist Republic
Bordering Countries: Afghanistan, Uzbekistan, Kyrgyzstan, China (4)
Administrative Divisions: Khatlon, Sughd (2 provinces) •Kuhistoni Badakhshon (1 autonomous province)
Ethnic/Racial Groups: Tajik, Uzbek, Russian
Religion: Islam (Sunni, Shiite)
Languages: Tajik, Russian
Currency: somoni
Current President: Emomali Rahmon
Cities (capital, largest, or with at least a million people):
  Dushanbe: •located in western Tajikistan •city in the Viloyati Khatlon province •capital of Tajikistan •most populated city in Tajikistan (704,000 people)

ENVIRONMENTAL/ECONOMIC:
Climate: Midlatitude continental with hot summers/mild winters; semi-arid to polar in the east
Natural Resources: hydropower, petroleum, uranium, mercury, brown coal
Agricultural Products: cotton, grain, fruits, grapes, vegetables
Major Exports: aluminum, electricity, cotton, fruits, vegetable oil, textiles
Natural Hazards: earthquakes, floods

TANZANIA
Country Name: United Republic of Tanzania
Continent: Africa
Area: 364,900 sq mi / 945,087 sq km
Population: 46,218,000
Population Density: 127 people per sq mi / 49 people per sq km
Capitals: Dar es Salaam (administrative capital), Dodoma (legislative capital)

PHYSICAL:
Highest Point: Kilimanjaro 19,341 ft / 5,895 m
Lowest Point: Indian Ocean 0 ft / 0 m (sea level)
Mountain Ranges:
  Pare: •located in northeastern Tanzania

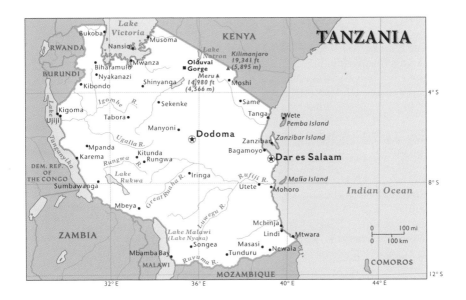

Peaks (minimum elevation of 8,000 ft / 2,400 m):
  Kilimanjaro: • located in the Pare Mountains • highest point in Africa
Valleys:
  Great Rift: • covers an area from northern to southern Tanzania • extends into Kenya, Burundi, Democratic Republic of the Congo, Zambia, and Malawi
Grasslands/Prairies:
  Serengeti Plain: • located in northern Tanzania • extends into Kenya
Oceans:
  Indian: • forms the eastern coast of Tanzania
Lakes:
  Lake Eyasi: • located in northern Tanzania • in the Great Rift Valley
  Lake Malawi: • forms most of Tanzania's border with Malawi • extends into Malawi and Mozambique
  Lake Natron: • located in northern Tanzania • in the Great Rift Valley • extends into Kenya
  Lake Rukwa: • located in western Tanzania • in the Great Rift Valley
  Lake Tanganyika: • located in western Tanzania • in the Great Rift Valley • on the border with Democratic Republic of the Congo, Burundi, and Zambia
  Lake Victoria: • located in northwestern Tanzania • extends into Uganda and Kenya • largest lake in Africa • third largest lake in the world
Islands:
  Zanzibar: • largest island off the coast of Tanzania • surrounded by the Indian Ocean

POLITICAL:
Independence: April 26, 1964 (from the United Kingdom)
Former Name: United Republic of Tanganyika and Zanzibar
Bordering Countries: Kenya, Mozambique, Malawi, Democratic Republic of the Congo, Burundi, Uganda, Zambia, Rwanda (8)
Administrative Divisions: Arusha, Dar es Salaam, Dodoma, Iringa, Kagera, Kigoma, Kilimanjaro, Lindi, Manyara, Mara, Mbeya, Morogoro, Mtwara, Mwanza, Pemba North, Pemba South, Pwani, Rukwa, Ruvuma, Shinyanga, Singida, Tabora, Tanga, Zanzibar Central/South, Zanzibar North, Zanzibar Urban/West (26 regions)
Ethnic/Racial Groups: predominantly Bantu on the mainland; Arab, African, mixed on Zanzibar
Religions: Christianity, Islam, indigenous beliefs
Languages: Kiswahili, Kiungujo, English, Arabic, many local languages
Currency: Tanzanian shilling
Current President: Jakaya Kikwete
Cities (capital, largest, or with at least a million people):
  Dar es Salaam: •located in eastern Tanzania •city in the Dar es Salaam region •administrative capital of Tanzania •most populated city in Tanzania (3,349,000 people) •chief port on the Indian Ocean
  Dodoma: •located in central Tanzania •city in the Dodoma region •legislative capital of Tanzania (200,000 people)

ENVIRONMENTAL/ECONOMIC:
Climate: tropical along the coast; temperate in the highlands
Natural Resources: hydropower, tin, phosphates, iron ore, coal
Agricultural Products: coffee, sisal, tea, cotton, cattle
Major Exports: gold, coffee, cashew nuts, manufactures, cotton
Natural Hazards: floods, droughts

THAILAND
Country Name: Kingdom of Thailand
Continent: Asia
Area: 198,115 sq mi / 513,115 sq km
Population: 69,519,000
Population Density: 351 people per sq mi / 135 people per sq km
Capital: Bangkok

PHYSICAL:
Highest Point: Doi Inthanon 8,415 ft / 2,565 m
Lowest Point: Gulf of Thailand 0 ft / 0 m (sea level)
Peaks (minimum elevation of 8,000 ft / 2,400 m):
  Doi Inthanon: •located in northwestern Thailand •highest point in Thailand
Plateau Regions:
  Khorat Plateau: •covers much of eastern Thailand

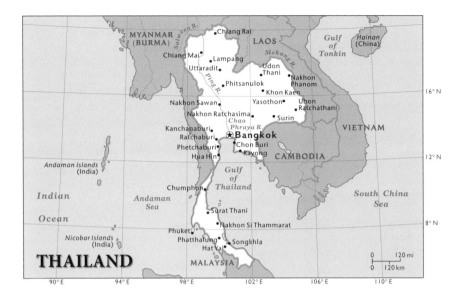

Seas:
Andaman: •forms much of the southwestern coast of Thailand •feeds into the Indian Ocean

Gulfs:
Gulf of Thailand: •forms the southeastern coast of Thailand •feeds into the South China Sea

Straits:
Strait of Malacca: •forms part of the southwestern coast of Thailand •separates the Malay Peninsula from Indonesia •connects the Andaman and South China Seas

Rivers:
Chao Phraya: •has its mouth in the Gulf of Thailand •has its source in northwestern Thailand

Mekong: •forms part of the Thailand-Laos border •extends into Laos, Cambodia, and Myanmar

Salween (Thanlwin): •forms part of the Thailand-Myanmar border •extends into Myanmar

Peninsulas:
Malay: •bordered by the Gulf of Thailand, South China Sea, Strait of Malacca, and Andaman Sea •forms the southernmost part of Thailand •includes part of Malaysia

POLITICAL:

Independence: 1238 (never colonized; traditional date for founding of the country)
Former Name: Siam
Bordering Countries: Myanmar, Laos, Cambodia, Malaysia (4)

Administrative Divisions: Amnat Charoen, Ang Thong, Buriram, Chachoengsao, Chai Nat, Chaiyaphum, Chanthaburi, Chiang Mai, Chiang Rai, Chon Buri, Chumphon, Kalasin, Kamphaeng Phet, Kanchanaburi, Khon Kaen, Krabi, Krung Thep Mahanakhon (Bangkok), Lampang, Lamphun, Loei, Lop Buri, Mae Hong Son, Maha Sarakham, Mukdahan, Nakhon Nayok, Nakhon Pathom, Nakhon Phanom, Nakhon Ratchasima, Nakhon Sawan, Nakhon Si Thammarat, Nan, Narathiwat, Nong Bua Lamphu, Nong Khai, Nonthaburi, Pathum Thani, Pattani, Phangnga, Phatthalung, Phayao, Phetchabun, Phetchaburi, Phichit, Phitsanulok, Phra Nakhon Si Ayutthaya, Phrae, Phuket, Prachin Buri, Prachuap Khiri Khan, Ranong, Ratchaburi, Rayong, Roi Et, Sa Kaeo, Sakon Nakhon, Samut Prakan, Samut Sakhon, Samut Songkhram, Sara Buri, Satun, Sing Buri, Sisaket, Songkhla, Sukhothai, Suphan Buri, Surat Thani, Surin, Tak, Trang, Trat, Ubon Ratchathani, Udon Thani, Uthai Thani, Uttaradit, Yala, Yasothon (76 provinces)
Ethnic/Racial Groups: Thai, Chinese
Religions: Buddhism, Islam
Languages: Thai, English, ethnic and regional dialects
Currency: baht
Current Prime Minister: Yingluck Shinawatra
Cities (capital, largest, or with at least a million people):
Bangkok: •located in central Thailand •city in the Krung Thep Mahanakhon (Bangkok) province •capital of Thailand •most populated city in Thailand (6,902,000 people) •chief port on the Chao Phraya River

ENVIRONMENTAL/ECONOMIC:
Climate: tropical with two monsoon seasons; hot and humid in the south
Natural Resources: tin, rubber, natural gas, tungsten, tantalum
Agricultural Products: rice, cassava (tapioca), rubber, corn
Major Exports: computers, transistors, seafood, clothing, rice
Natural Hazards: land subsidence, droughts

TIMOR-LESTE see East Timor

TOGO
Country Name: Republic of Togo
Continent: Africa
Area: 21,925 sq mi / 56,785 sq km
Population: 5,847,000
Population Density: 267 people per sq mi / 103 people per sq km
Capital: Lome

PHYSICAL:
Highest Point: Mont Agou 3,235 ft / 986 m
Lowest Point: Bight of Benin 0 ft / 0 m (sea level)
Bays:
Bight of Benin: •forms the southern coast of Togo •feeds into the Gulf of Guinea

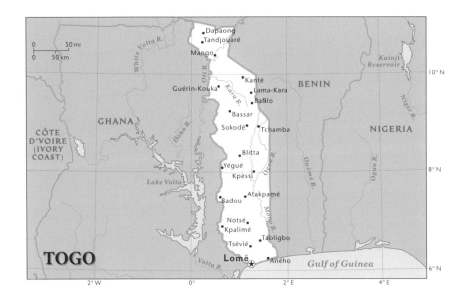

POLITICAL:
Independence: April 27, 1960 (from French administration)
Former Name: French Togoland
Bordering Countries: Ghana, Benin, Burkina Faso (3)
Administrative Divisions: Centrale, Kara, Maritime, Plateaux, Savanes (5 regions)
Ethnic/Racial Groups: Ewe, Mina, Kabre, and other African tribes
Religions: indigenous beliefs, Christianity, Islam
Languages: French, Ewe, Mina, Kabye, Dagomba
Currency: Communaute Financiere Africaine franc
Current President: Faure Gnassingbe
Cities (capital, largest, or with at least a million people):
  Lome: •located in southern Togo •city in the Maritime region •capital of Togo •most populated city in Togo (1,593,000 people) •chief port on the Bight of Benin

ENVIRONMENTAL/ECONOMIC:
Climate: tropical; hot and humid in the south; semi-arid in the north
Natural Resources: phosphates, limestone, marble, arable land
Agricultural Products: coffee, cacao, cotton, yams, livestock
Major Exports: re-exports, cotton, phosphates, coffee, cacao
Natural Hazards: harmattan winds, droughts

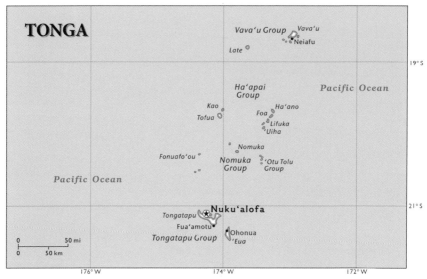

TONGA

Country Name: Kingdom of Tonga
Continent: Australia/Oceania
Area: 289 sq mi / 748 sq km
Population: 104,000
Population Density: 360 people per sq mi / 139 people per sq km
Capital: Nukualofa

PHYSICAL:
Highest Point: unnamed location 3,389 ft / 1,033 m
Lowest Point: Pacific Ocean 0 ft / 0 m (sea level)
Oceans:
  Pacific: •surrounds Tonga
Islands:
  Tongatapu: •largest island in Tonga •surrounded by the Pacific Ocean

POLITICAL:
Independence: June 4, 1970 (from the United Kingdom)
Former Name: Friendly Islands
Administrative Divisions: Haapai, Tongatapu, Vavau (3 island groups)
Ethnic/Racial Groups: Polynesian
Religion: Christianity
Languages: Tongan, English
Currency: paanga
Current King: George Tupou V

Cities (capital, largest, or with at least a million people):
  Nukualofa: •located on the island of Tongatapu •city in the Tongatapu island group •capital of Tonga •most populated city in Tonga (24,000 people) •chief port on the Pacific Ocean

ENVIRONMENTAL/ECONOMIC:
Climate: tropical, modified by trade winds with warm/cool seasons
Natural Resources: fish, fertile soil
Agricultural Products: squash, coconuts, copra, bananas, fish
Major Exports: squash, fish, vanilla beans, root crops
Natural Hazards: cyclones, earthquakes, volcanoes

TRINIDAD AND TOBAGO
Country Name: Republic of Trinidad and Tobago
Continent: North America
Area: 1,980 sq mi / 5,128 sq km
Population: 1,325,000
Population Density: 669 people per sq mi / 258 people per sq km
Capital: Port-of-Spain

PHYSICAL:
Highest Point: El Cerro del Aripo 3,084 ft / 940 m
Lowest Point: Atlantic Ocean 0 ft / 0 m (sea level)
Oceans:
  Atlantic: •forms the eastern and southern coast of Trinidad and Tobago
Seas:
  Caribbean: •forms the northern coast of Trinidad •forms the western and southern coasts of Tobago •feeds into the Atlantic Ocean
Gulfs:
  Gulf of Paria: •forms the western coast of Trinidad •separates Trinidad and Venezuela
Islands:
  Tobago: •second largest island in Trinidad and Tobago •bordered by the Atlantic Ocean and Caribbean Sea
  Trinidad: •largest island in Trinidad and Tobago •bordered by the Atlantic Ocean, Caribbean Sea, and Gulf of Paria

POLITICAL:
Independence: August 31, 1962 (from the United Kingdom)
Administrative Divisions: Caroni, Mayaro, Nariva, Saint Andrew, Saint David, Saint George, Saint Patrick, Victoria (8 counties) •Arima, Port-of-Spain, San Fernando (3 municipalities) • Tobago (1 ward)
Ethnic/Racial Groups: black, East Indian, mixed
Religions: Christianity (Roman Catholic, Anglican), Hindu

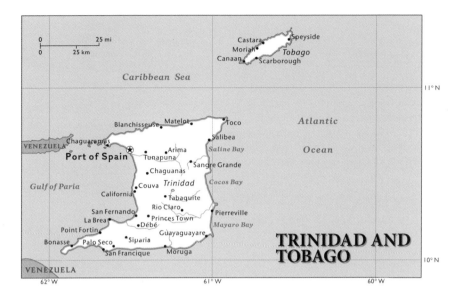

Languages: English, Hindi, French, Spanish
Currency: Trinidad and Tobago dollar
Current President: George Maxwell Richards
Cities (capital, largest, or with at least a million people):
  Port-of-Spain: •located on the island of Trinidad •city in the Port-of-Spain municipality •capital of Trinidad and Tobago •most populated city in Trinidad and Tobago (57,000 people) •chief port on the Gulf of Paria

ENVIRONMENTAL/ECONOMIC:
Climate: Tropical with a rainy season
Natural Resources: petroleum, natural gas
Agricultural Products: cacao, sugarcane, rice, citrus
Major Exports: petroleum and petroleum products, chemicals, steel
Natural Hazards: tropical storms

TUNISIA
Country Name: Republic of Tunisia
Continent: Africa
Area: 63,170 sq mi / 163,610 sq km
Population: 10,676,000
Population Density: 169 people per sq mi / 65 people per sq km
Capital: Tunis

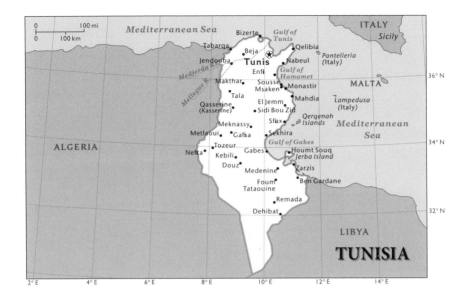

PHYSICAL:
Highest Point: Jebel ech Chambi 5,066 ft / 1,544 m
Lowest Point: Shott el Gharsa 56 ft / 17 m below sea level
Mountain Ranges:
  Atlas: •located in northwestern Tunisia •extend into Algeria
Deserts:
  Grand Erg Oriental: •covers much of western Tunisia •extends into Algeria
  Sahara: •covers much of southern and central Tunisia •extends into Algeria and Libya •largest desert in the world
Seas:
  Mediterranean: •forms much of the northern and eastern coasts of Tunisia •feeds into the Atlantic Ocean
Gulfs:
  Gulf of Gabes: •forms part of the eastern coast of Tunisia •feeds into the Mediterranean Sea
  Gulf of Tunis: •forms much of the northeastern coast of Tunisia •feeds into the Mediterranean Sea
Lakes:
  Shott el Jerid: •located in western Tunisia •on the edge of the Grand Erg Oriental

POLITICAL:
Independence: March 20, 1956 (from France)
Bordering Countries: Algeria, Libya (2)

Administrative Divisions: Al Kef, Al Mahdiyah, Al Munastir, Al Qasrayn, Al Qayrawan, Ariana, Bajah, Banzart, Bin Arus, Jundubah, Madanin, Manouba, Nabul, Qabis, Qafsah, Qibili, Safaqis, Sidi Bu Zayd, Silyanah, Susah, Tatawin, Tawzar, Tunis, Zaghwan (24 governorates)
Ethnic/Racial Groups: Arab, European
Religion: predominantly Islam
Languages: Arabic, French
Currency: Tunisian dinar
Current President: Faoud M'Bazaa
Cities (capital, largest, or with at least a million people):
    Tunis: •located in northern Tunisia •city in the Tunis governorate •capital of Tunisia •most populated city in Tunisia (759,000 people) •chief port on the Gulf of Tunis

ENVIRONMENTAL/ECONOMIC:
Climate: temperate in the north with mild and rainy winters/hot and dry summers; desert in the south
Natural Resources: petroleum, phosphates, iron ore, lead, zinc
Agricultural Products: olives, olive oil, grain, dairy products
Major Exports: textiles, mechanical goods, phosphates and chemicals, agricultural products
Natural Hazards: droughts

TURKEY
Country Name: Republic of Turkey
Continent: Asia/Europe
Area: 300,948 sq mi / 779,452 sq km
Population: 73,950,000
Population Density: 246 people per sq mi / 95 people per sq km
Capital: Ankara

PHYSICAL:
Highest Point: Mount Ararat 16,853 ft / 5,137 m
Lowest Point: Mediterranean Sea 0 ft / 0 m (sea level)
Mountain Ranges:
    Lesser Caucasus: •located in northeastern Turkey •extend into Georgia and Armenia
    Taurus: •cover an area from southern to eastern Turkey
    Zagros: •cover much of eastern Turkey •extend into Iran and Iraq
Peaks (minimum elevation of 8,000 ft / 2,400 m):
    Mount Ararat: •located in the Zagros Mountains •highest point in Turkey
Seas:
    Aegean: •forms part of the western coast of Turkey •feeds into the Mediterranean Sea and Sea of Crete

Black: •forms most of the northern coast of Turkey •feeds into the Sea of Marmara
Mediterranean: •forms most of the southern coast of Turkey •feeds into the Atlantic Ocean
Sea of Marmara: •forms part of the northwestern coast of Turkey •feeds into the Aegean Sea

Gulfs:
Gulf of Antalya: •forms part of the southwestern coast of Turkey •feeds into the Mediterranean Sea

Straits:
Bosporus: •forms part of the northwestern coast of Turkey •separates European Turkey from Asian Turkey •connects the Black Sea and Sea of Marmara
Dardanelles: •forms part of the northwestern coast of Turkey •separates European Turkey from Asian Turkey •connects the Sea of Marmara and Aegean Sea

Lakes:
Ataturk Reservoir: •located in southern Turkey •on the edge of the Taurus Mountains •fed by the Euphrates River •formed by the Ataturk Dam
Lake Tuz: •located in central Turkey
Lake Van: •located in eastern Turkey •on the edge of the Zagros Mountains

Rivers:
Aras: •has its source in the Taurus Mountains •forms part of the Turkey-Armenia border
Euphrates: •has its source in the Zagros Mountains •extends into Syria and Iraq
Evros: •has its mouth in the Aegean Sea •forms the Turkey-Greece border
Orontes: •has its mouth in the Mediterranean Sea •extends into Syria
Tigris: •has its source in the Taurus Mountains •extends into Iraq

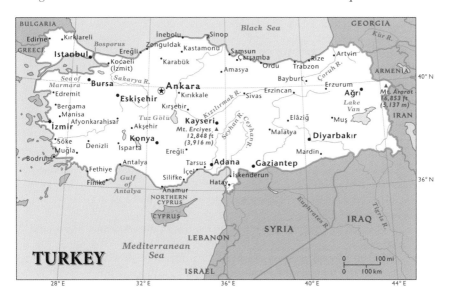

COUNTRIES A TO Z | PAGE 341

Dams:
>Ataturk: •located in southern Turkey •on the Euphrates River •forms the Ataturk Reservoir

Regions:
>Anatolia: •mountainous region that covers most of Turkey •includes the Taurus Mountains and much of the Zagros Mountains •east of the Aegean Sea and Sea of Marmara •south of the Black Sea •north of the Mediterranean Sea and the Gulf of Antalya

POLITICAL:
Independence: October 29, 1923 (from the Ottoman Empire)
Bordering Countries: Syria, Iran, Iraq, Armenia, Georgia, Bulgaria, Greece, Azerbaijan (8)
Regions:
>Cappadocia: •region that covers much of central Turkey •includes part of the Taurus Mountains
>Kurdistan: •region in southeastern Turkey •south of the Taurus Mountains •includes part of the Zagros Mountains •extends into Iran and Iraq
>Thrace: •region that covers most of European Turkey •south of the Black Sea and north of the Sea of Marmara •extends into Bulgaria and Greece

Administrative Divisions: Adana, Adiyaman, Afyonkarahisar, Agri, Aksaray, Amasya, Ankara, Antalya, Ardahan, Artvin, Aydin, Balikesir, Bartin, Batman, Bayburt, Bilecik, Bingol, Bitlis, Bolu, Burdur, Bursa, Canakkale, Cankiri, Corum, Denizli, Diyarbakir, Duzce, Edirne, Elazig, Erzincan, Erzurum, Eskisehir, Gaziantep, Giresun, Gumushane, Hakkari, Hatay, Igdir, Isparta, Istanbul, Izmir, Kahramanmaras, Karabuk, Karaman, Kars, Kastamonu, Kayseri, Kilis, Kirikkale, Kirklareli, Kirsehir, Kocaeli, Konya, Kutahya, Malatya, Manisa, Mardin, Mersin, Mugla, Mus, Nevsehir, Nigde, Ordu, Osmaniye, Rize, Sakarya, Samsun, Sanliurfa, Siirt, Sinop, Sirnak, Sivas, Tekirdag, Tokat, Trabzon, Tunceli, Usak, Van, Yalova, Yozgat, Zonguldak (81 provinces)
Ethnic/Racial Groups: Turkish, Kurdish
Religion: predominantly Islam
Languages: Turkish, Kurdish, Arabic, Armenian, Greek
Currency: Turkish lira
Current President: Abdullah Gul
Cities (capital, largest, or with at least a million people):
>Istanbul: •located in northwestern Turkey •city in the Istanbul province •most populated city in Turkey (10,525,000 people) •chief port on the Sea of Marmara and the Bosporus Strait
>Ankara: •located in central Turkey •city in the Ankara province •capital of Turkey (3,846,000 people)
>Izmir: •located in western Turkey •city in the Izmir province •chief port on the Aegean Sea
>Bursa: •located in northwestern Turkey •city in the Bursa province
>Adana: •located in southern Turkey •city in the Adana province •on the edge of the

Taurus Mountains
Gaziantep: •located in southern Turkey •city in the Gaziantep province

ENVIRONMENTAL/ECONOMIC:
Climate: temperate with hot, dry summers/mild, wet winters; harsher in the mountains
Natural Resources: coal, iron ore, copper, chromium, antimony
Agricultural Products: tobacco, cotton, grain, olives, livestock
Major Exports: apparel, foodstuffs, textiles, metal manufactures, transport equipment
Natural Hazards: earthquakes

TURKMENISTAN
Country Name: Turkmenistan
Continent: Asia
Area: 188,456 sq mi / 488,100 sq km
Population: 5,105,000
Population Density: 27 people per sq mi / 10 people per sq km
Capital: Ashgabat

PHYSICAL:
Highest Point: Gora Ayribaba 10,299 ft / 3,139 m
Lowest Point: Vpadina Akchanaya 266 ft / 81 m below sea level
Peaks (minimum elevation of 8,000 ft / 2,400 m):

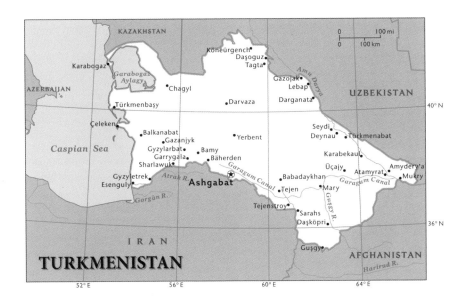

Gora Ayribaba: •located in eastern Turkmenistan •on the Turkmenistan-Uzbekistan border •highest point in Turkmenistan
Depressions:
Vpadina Akchanaya: •located in northern Turkmenistan •lowest point in Turkmenistan
Caspian Sea: •forms the western coast of Turkmenistan
Desert:
Garagum: •covers most of Turkmenistan
Lakes:
Caspian Sea: •forms most of western coast of Turkmenistan •extends into Kazakhstan and Iran •largest lake in the world
Rivers:
Amu Darya: •forms part of Turkmenistan's border with Uzbekistan and Afghanistan •extends into Uzbekistan

POLITICAL:
Independence: October 27, 1991 (from the Soviet Union)
Former Name: Turkmen Soviet Socialist Republic
Bordering Countries: Uzbekistan, Iran, Afghanistan, Kazakhstan (4)
Administrative Divisions: Ahal, Balkan, Dasoguz, Lebap, Mary (5 provinces)
Ethnic/Racial Groups: Turkmen, Uzbek, Russian, Kazakh
Religions: Islam, Christianity (primarily Eastern Orthodox)
Languages: Turkmen, Russian, Uzbek
Currency: Turkmen manat
Current President: Gurbanguly Berdimuhamedow
Cities (capital, largest, or with at least a million people):
Ashgabat: •located in southen Turkmenistan •city in the Ahal province •capital of Turkmenistan •most populated city in Turkmenistan (637,000 people) •on the edge of the Garagum desert

ENVIRONMENTAL/ECONOMIC:
Climate: subtropical desert
Natural Resources: petroleum, natural gas, coal, sulfur, salt
Agricultural Products: cotton, grains, livestock
Major Exports: gas, oil, cotton fiber, textiles
Natural Hazards: droughts, sandstorms

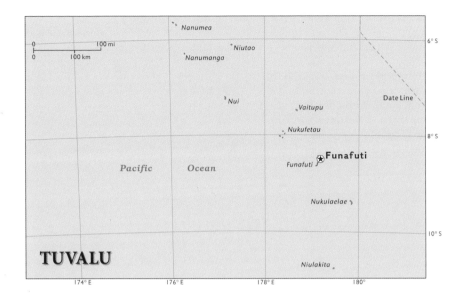

TUVALU

Country Name: Tuvalu
Continent: Australia/Oceania
Area: 10 sq mi / 26 sq km
Population: 11,000 (smallest in Australia/Oceania)
Population Density: 1,100 people per sq mi / 423 people per sq km
Capital: Funafuti

PHYSICAL:

Highest Point: unnamed location 16 ft / 5 m
Lowest Point: Pacific Ocean 0 ft / 0 m (sea level)
Oceans:
   Pacific: surrounds Tuvalu

POLITICAL:

Independence: October 1, 1978 (from the United Kingdom)
Former Name: Ellice Islands
Ethnic/Racial Groups: Polynesian, Micronesian
Religion: Christianity; predominantly Church of Tuvalu (Congregationalist)
Languages: Tuvaluan, English, Samoan
Currency: Australian dollar, Tuvaluan dollar

Current Prime Minister: Willie Telavi

Cities (capital, largest, or with at least a million people):
    Funafuti: •located on an island in central Tuvalu •capital of Tuvalu •most populated city in Tuvalu (5,000 people) •chief port on the Pacific Ocean

ENVIRONMENTAL/ECONOMIC:
Climate: tropical, moderated by easterly trade winds; heavy rains in winter
Natural Resources: fish
Agricultural Products: coconuts, fish
Major Exports: copra, fish
Natural Hazards: cyclones, high sea levels

UGANDA

Country Name: Republic of Uganda
Continent: Africa
Area: 93,104 sq mi / 241,139 sq km
Population: 34,543,000
Population Density: 371 people per sq mi / 143 people per sq km
Capital: Kampala

PHYSICAL:

Highest Point: Margherita Peak (on Mount Stanley) 16,765 ft / 5,110 m
Lowest Point: Lake Albert 2,037 ft / 621 m
Mountain Ranges:
  Bleus: •cover much of western Uganda •extend into Democratic Republic of the Congo
  Mitumba: •located in southwestern Uganda •extend into Democratic Republic of the Congo and Rwanda
Peaks (minimum elevation of 8,000 ft / 2,400 m):
  Mount Stanley: •located in the Bleus Mountains •on the Uganda-Democratic Republic of the Congo border •highest point in Uganda
Lakes:
  Lake Albert: •located in western Uganda •on the edge of the Bleus Mountains •fed by the Victoria Nile River •on the Uganda–Democratic Republic of the Congo border
  Lake Edward: •located in southwestern Uganda •on the edge of the Bleus Mountains •on the Uganda–Democratic Republic of the Congo border
  Lake Kyoga: •located in central Uganda •fed by the Victoria Nile River
  Lake Victoria: •covers much of southern Uganda •Owens Falls Dam on the Victoria Nile River makes the lake a reservoir •extends into Tanzania and Kenya •largest lake in Africa •third largest lake in the world
Rivers:
  Albert Nile: •extends into Sudan
  Victoria Nile: •tributary of the Albert Nile River •flows into Lake Albert •has its source in Lake Victoria •Owens Falls Dam creates hydroelectric power from the river
Dams:
  Owen Falls: •located in southeastern Uganda •crosses the Victoria Nile River •makes Lake Victoria a reservoir

POLITICAL:

Independence: October 9, 1962 (from the United Kingdom)

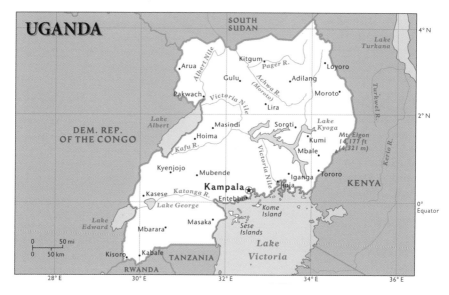

Bordering Countries: Kenya, Democratic Republic of the Congo, South Sudan, Tanzania, Rwanda (5)
Administrative Divisions: Uganda has more than 100 districts, so they are not listed here. Please go to CIA.gov to see the full list.
Ethnic/Racial Groups: Baganda, Ankole, Basoga, Itesco, Bakiga, other indigenous groups
Religions: Christianity (Roman Catholic, Protestant), indigenous beliefs, Islam
Languages: English, Ganda or Luganda, many local languages
Currency: Ugandan shilling
Current President: Yoweri Kaguta Museveni
Cities (capital, largest, or with at least a million people):
  Kampala: •located in southern Uganda •city in the Kampala district •capital of Uganda •most populated city in Uganda (1,535,000 people) •chief port on Lake Victoria

ENVIRONMENTAL/ECONOMIC:
Climate: tropical with rainy/dry seasons; semi-arid in the northeast
Natural Resources: copper, cobalt, hydropower, limestone, salt
Agricultural Products: coffee, tea, cotton, tobacco, beef
Major Exports: coffee, fish and fish products, tea, gold, cotton
Natural Hazards: floods, lalndslides, thunderstorms

UKRAINE
Country Name: Ukraine
Continent: Europe
Area: 233,090 sq mi / 603,700 sq km (largest country completely in Europe)

Population: 45,730,000
Population Density: 196 people per sq mi / 76 people per sq km
Capital: Kiev

PHYSICAL:
Highest Point: Hora Hoverla 6,762 ft / 2,061 m
Lowest Point: Black Sea 0 ft / 0 m (sea level)
Mountain Ranges:
  Carpathian: •located in western Ukraine •extend into Poland, Slovakia, and Romania
Grasslands/Prairies:
  Great Alfold: •located in western Ukraine •extends into Slovakia, Hungary, and Romania
  Black Sea Lowland: •covers most of southern Ukraine •extends into Moldova and Romania
Swamps:
  Pinsk Marshes: •located in northwestern Ukraine •extend into Belarus
Seas:
  Black: •forms part of the southern coast of Ukraine •feeds into the Sea of Marmara
  Sea of Azov: •forms part of the southeastern coast of Ukraine •feeds into the Black Sea
Rivers:
  Danube: •has its mouth in the Black Sea •forms part of the Ukraine-Romania border •extends into Romania and Hungary
  Dnieper: •has its mouth in the Black Sea •forms part of the Ukraine-Belarus border •extends into Belarus and Russia
  Dniester (Nistru): •has its mouth in the Black Sea •has its source in the Carpathian

Mountains •forms part of the Ukraine-Moldova border •extends into Moldova
Donets: •extends into Russia
Prut: •has its source in the Carpathian Mountains •forms part of the Ukraine-Romania border
Tisza: •has its source in the Carpathian Mountains •forms part of the Ukraine-Romania border •extends into Hungary

Peninsulas:
Crimean: •bordered by the Black Sea and Sea of Azov

POLITICAL:
Independence: August 24, 1991 (from the Soviet Union)
Former Names: Ukrainian National Republic, Ukrainian State, Ukrainian Soviet Socialist Republic
Bordering Countries: Russia, Moldova, Belarus, Romania, Poland, Hungary, Slovakia (7)
Regions:
Bessarabia: •region in southwestern Ukraine •includes part of the Black Sea Lowland •northwest of the Black Sea •extends into Moldova
Administrative Divisions: Cherkaska, Chernihivska, Chernivetska, Dnipropetrovska, Donetska, Ivano-Frankivska, Kharkivska, Khersonska, Khmelnytska, Kirovohradska, Kyyivska, Luhanska, Lvivska, Mykolayivska, Odeska, Poltavska, Rivnenska, Sumska, Ternopilska, Vinnytska, Volynska, Zakarpatska, Zaporizka, Zhytomyrska (24 oblasti) •Avtonomna Respublika Krym (1 autonomous republic) •Kiev, Sevastopol (2 municipalities)
Ethnic/Racial Groups: Ukrainian, Russian
Religions: Christianity (Ukrainian Orthodox, Ukrainian Catholic, Protestant), Judaism
Languages: Ukrainian, Russian, Romanian, Polish, Hungarian
Currency: hryvnia
Current President: Viktor Yanukovych
Cities (capital, largest, or with at least a million people):
Kiev: •located in northern Ukraine •city in the Kiev municipality •capital of Ukraine •most populated city in Ukraine (2,779,000 people) •chief port on the Dnieper River
Kharkiv: •located in northeastern Ukraine •city in the Kharkivska oblast
Odesa: •located in southwestern Ukraine •city in the Odeska oblast •on the edge of the Black Sea Lowland •chief port on the Black Sea
Dnipropetrovsk: •located in eastern Ukraine •city in the Dnipropetrovs'ka oblast •a major port on the Dnieper River

ENVIRONMENTAL/ECONOMIC:
Climate: temperate continental; Mediterranean on the south Crimean coast, wettest in the north and west, cool/cold winters, warm/hot summers
Natural Resources: iron ore, coal, manganese, natural gas, oil
Agricultural Products: grain, sugar beets, sunflower seed, beef
Major Exports: ferrous and nonferrous metals, fuel and petroleum products,

chemicals, machinery and transport equipment
Natural Hazards: floods, landslides, hail, snowstorms

UNITED ARAB EMIRATES
Country Name: United Arab Emirates
Continent: Asia
Area: 30,000 sq mi / 77,700 sq km
Population: 7,891,000
Population Density: 263 people per sq mi / 102 people per sq km
Capital: Abu Dhabi

PHYSICAL:
Highest Point: Jabal Yibir 5,010 ft / 1,527 m
Lowest Point: Persian Gulf 0 ft / 0 m (sea level)
Deserts:
   Rub al Khali: •covers most of the country •extends into Saudi Arabia and Oman
Gulfs:
   Gulf of Oman: •forms part of the eastern coast of United Arab Emirates •feeds into the Arabian Sea
   Persian: •forms most of the coast of United Arab Emirates •feeds into the Gulf of Oman

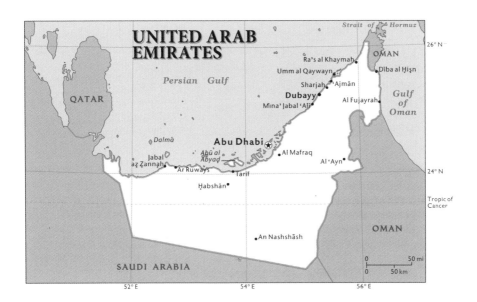

POLITICAL:
Independence: December 2, 1971 (from the United Kingdom)
Former Names: Trucial Oman, Trucial States
Bordering Countries: Saudi Arabia, Oman (2)
Administrative Divisions: Abu Dhabi, Ajman, Al Fujayrah, Ash Shariqah, Dubayy, Ras al Khaymah, Umm al Qaywayn (7 emirates)
Ethnic/Racial Groups: Arab, Iranian, Indian
Religions: Islam (Sunni and Shiite), Christianity, Hinduism
Languages: Arabic, Persian, English, Hindi, Urdu
Currency: Emirati dirham
Current President: Khalifa bin Zayid al-Nuhayyan
Cities (capital or with at least a million people):
  Dubayy: •located in northeastern United Arab Emirates •city in the Dubayy emirate •most populated city in United Arab Emirates (1,567,000 people) •on the edge of the Rub al Khali •chief port on the Persian Gulf
  Abu Dhabi: •located on an island in central United Arab Emirates •city in the Abu Dhabi emirate •capital of the United Arab Emirates (666,000 people) •a major port on the Persian Gulf

ENVIRONMENTAL/ECONOMIC:
Climate: desert; cooler in the mountains
Natural Resources: petroleum, natural gas
Agricultural Products: dates, vegetables, watermelons, poultry, fish
Major Exports: crude oil, natural gas, re-exports, dried fish, dates
Natural Hazards: sandstorms, dust storms

UNITED KINGDOM

Country Name: United Kingdom of Great Britain and Northern Ireland
Continent: Europe
Area: 93,788 sq mi / 242,910 sq km
Population: 62,588,000
Population Density: 667 people per sq mi / 258 people per sq km
Capital: London

PHYSICAL:
Highest Point: Ben Nevis 4,406 ft / 1,343 m
Lowest Point: The Fens 13 ft / 4 m below sea level
Mountain Ranges:
  Cambrian: •located in central Wales on the island of Great Britain
  Grampian Hills: •located in Scotland on the island of Great Britain •forms a boundary between the Scottish Highlands and the Scottish Lowlands •includes Ben Nevis, highest peak in the United Kingdom
  Pennines: •located on the island of Great Britain •extend from the Scotland-England border into central England

Valleys:
   Glen Mor: •located on the island of Great Britain •on the edge of the Grampian Hills
Grasslands/Prairies:
   Salisbury Plain: •located in southwestern England •on the island of Great Britain •site of Stonehenge
Oceans:
   Atlantic: •forms part of the western and northern coasts of the United Kingdom
Seas:
   Celtic: •forms part of the southwestern coast of the United Kingdom •feeds into the Atlantic Ocean
   Irish: •forms part of the inner coast of the United Kingdom •feeds into the Celtic Sea and Atlantic Ocean
   North: •forms most of the eastern coast of the United Kingdom •feeds into the Atlantic Ocean
   Sea of the Hebrides: •forms part of the northwestern coast of the United Kingdom •feeds into the Atlantic Ocean
Bays:
   Cardigan Bay: •forms part of the western coast of Great Britain •feeds into the Irish Sea
   Firth of Clyde: •forms part of the western coast of the island of Great Britain •feeds into the North Channel
   Firth of Forth: •forms part of the northeastern coast of the island of Great Britain •feeds into the North Sea

    Moray Firth: •forms part of the northern coast of the island of Great Britain •feeds into the North Sea

Straits:
    Strait of Dover: •forms part of the southeastern coast of the island of Great Britain •separates the island of Great Britain from France •connects the English Channel and the North Sea

Channels:
    Bristol: •forms part of the southwestern coast of the island of Great Britain •feeds into the Celtic Sea
    English: •forms much of the southern coast of the island of Great Britain •separates the island of Great Britain and France •connects the North Sea and the Atlantic Ocean
    North: •forms part of the inner coast of the United Kingdom •separates the island of Great Britain from the island of Ireland •connects the Irish Sea and the Atlantic Ocean
    Saint George's: •forms part of the southwestern coast of the island of Great Britain •separates the island of Great Britain and Ireland •connects the Irish and Celtic Seas

Passages:
    The Minch: •forms part of the northwestern coast of the United Kingdom •separates the Outer Hebrides from the island of Great Britain •connects the Sea of the Hebrides and Atlantic Ocean

Lakes:
    Loch Lomond: •located on the island of Great Britain •on the edge of the Grampian Hills
    Lough Foyle: •located in northern Northern Ireland •on the United Kingdom-Ireland border
    Lough Neagh: •located in central Northern Ireland

Rivers:
    Mersey: •has its mouth in the Irish Sea •has its source in the Pennines
    Severn: •has its mouth in the Bristol Channel •has its source in the Cambrian Mountains
    Thames: •has its mouth in the North Sea •has its source in southern England
    Trent: •has its mouth in the Humber Estuary •has its source in the Pennines

Deltas:
    Humber Estuary: •mouth of the Trent River •feeds into the North Sea

Capes:
    Land's End: •bordered by the Celtic Sea, Atlantic Ocean, and English Channel •southwesternmost point in the United Kingdom

Islands:
    Great Britain: •largest island in the United Kingdom •bordered by the Atlantic Ocean, Sea of the Hebrides, North Sea, Irish Sea, Celtic Sea, Moray Firth, Firth of Clyde, Firth of Forth, Cardigan Bay, Strait of Dover, North Channel, Saint George's Channel, Bristol Channel, English Channel, and The Minch
    Inner Hebrides: •island group off the northwestern coast of the United Kingdom

- bordered by the Atlantic Ocean and Sea of the Hebrides

Ireland: •second largest island in the United Kingdom •bordered by the Atlantic Ocean, Irish Sea, North Channel, and St. George's Channel •made up of the country of Ireland and Northern Ireland (part of the United Kingdom)

Isle of Lewis: •largest island in the Outer Hebrides •bordered by the Atlantic Ocean and The Minch

Orkney: •island group off the northern coast of the United Kingdom •bordered by the Atlantic Ocean and North Sea

Outer Hebrides: •island group off the northwestern coast of the United Kingdom •bordered by the Atlantic Ocean and The Minch

Shetland: •island group north of the Orkney Islands •bordered by the Atlantic Ocean and North Sea

POLITICAL:

Independence: 10th century (England), 1536 (union of England and Wales formalized), 1707 (union of England and Scotland as Great Britain), 1801 (union of Great Britain and Ireland), 1927 (present name adopted)

Bordering Country: Ireland (1)

Regions:
 Ulster: •region that covers Northern Ireland •extends into Ireland

Administrative Divisions: England, Northern Ireland, Scotland, Wales (4 countries)

External Territories: Anguilla, Bermuda, British Indian Ocean Territory, British Virgin Islands, Cayman Islands, Channel Islands, Falkland Islands, Gibraltar, Isle of Man, Montserrat, Pitcairn Islands, Saint Helena, South Georgia and the South Sandwich Islands, Turks and Caicos Islands (14)

Ethnic/Racial Groups: English, Scottish, Irish, Welsh, Ulster

Religions: Christianity (Anglican, Roman Catholic, other Protestant), Islam

Languages: English, Welsh, Scottish form of Gaelic

Currency: British pound

Current Prime Minister: David Cameron

Cities (capital, largest, or with at least a million people):
 London: •located on the island of Great Britain •city in southeastern England •capital of the United Kingdom •most populated city in the United Kingdom (8,615,000 people) •chief port on the River Thames

 Manchester-Liverpool: •located on the island of Great Britain •metropolitan area in central England

 Glasgow: •located on the island of Great Britain •city in western Scotland

 Birmingham: •located on the island of Great Britain •city in central England

ENVIRONMENTAL/ECONOMIC:

Climate: temperate, moderated by the North Atlantic Current

Natural Resources: coal, petroleum, natural gas, iron ore, lead

Agricultural Products: cereals, oilseed, potatoes, vegetables, cattle

Major Exports: manufactured goods, fuels, chemicals, food, beverages

Natural Hazards: winter windstorms, floods

UNITED STATES

Country Name: United States of America
Continent: North America
Area: 3,794,083 sq mi / 9,826,630 sq km
Population: 311,695,000 (largest in North America)
Population Density: 82 people per sq mi / 32 people per sq km
Capital: Washington, D.C.

PHYSICAL:
Highest Point: Mount McKinley 20,320 ft / 6,194 m
Lowest Point: Death Valley 282 ft / 86 m below sea level
Mountain Ranges:
  Alaska: •located in southern Alaska •includes Mount McKinley (Denali) highest point in North America
  Aleutian: •located in southwestern Alaska on the Aleutian Peninsula
  Appalachian: •stretch north-south across the eastern United States •extend into Canada
  Brooks: •located in northern Alaska
  Cascade: •located near the Pacific Coast •east of the Coast Ranges •extends from northern California to Washington •has many volcanic peaks •includes Mount St. Helens
  Coast: •located along the western coast of the contiguous United States •extend from California to Washington •continue along the southern coast of Alaska

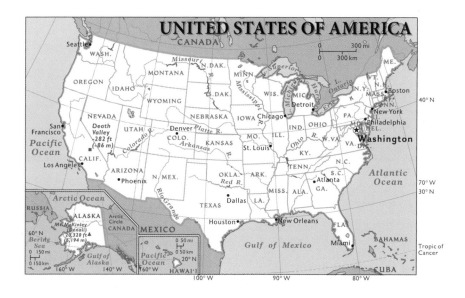

Rocky : •cover much of the western United States •just west of the Great Plains •extend into Canada •form a continental divide
Saint Elias: •located in southeastern Alaska •extend from Alaska into Canada
Sierra Nevada: •located in eastern California in the western United States

Peaks (minimum elevation of 8,000 ft / 2,400 m):
Mount McKinley: •located in the Alaska Range •highest point in North America •also called Denali
Mount Whitney: •located in the Sierra Nevada •highest point in the contiguous United States

Passes:
Chilkoot: •located in the Coast Mountains of Alaska •crosses the border into Canada

Valleys:
San Joaquin: •located in California between the Coast Ranges and the Sierra Nevada

Grasslands/Prairies:
Great Plains: •cover much of the central United States •located between the Rocky Mountains and the Mississippi River •extend into Canada
Coastal Plain: •extends from the Gulf Coast of the United States northeast along the Atlantic Coast

Plateau Regions:
Colorado Plateau: •located in the southwestern United States •between the Great Basin and the Rocky Mountains
Columbia Plateau: •located in the northwestern United States •between the Cascade Range and Rocky Mountains
Great Basin: •covers much of the western United States •desert region between the Sierra Nevada and Rocky Mountains
Ozark Plateau: •forested plateau in the south-central United States •covers much of Arkansas •extends into Missouri

Depressions:
Death Valley: •located in southwestern United States •in California near the Nevada border •part of the Mojave Desert •lowest point in North America

Swamps:
The Everglades: •located in southeastern United States •cover much of the southern part of the Florida Peninsula

Deserts:
Chihuahuan: •located in the southwestern United States •in Texas and New Mexico •extends into Mexico
Mojave: •located in the southwestern United States •primarily in southern California and Nevada
Sonoran: •located in the southwestern United States •primarily in Arizona •extends into Mexico

Oceans:
Arctic: •forms much of the northern coast of Alaska
Atlantic: •forms most of the eastern coast of the United States
Pacific: •forms most of the western coast of the United States

Seas:
    Beaufort: •forms part of the northeastern coast of Alaska •feeds into the Arctic Ocean
    Bering: •forms part of the western coast of Alaska •feeds into the Pacific Ocean
    Chukchi: •forms part of the northwestern coast of Alaska •feeds into the Arctic Ocean

Gulfs:
    Gulf of Alaska: •forms part of the southern coast of Alaska •feeds into the Pacific Ocean
    Gulf of Maine: •forms part of the northeastern coast of the United States •feeds into the Atlantic Ocean
    Gulf of Mexico: •forms much of the southern coast of the United States •feeds into the Caribbean Sea and Atlantic Ocean

Bays:
    Chesapeake: •forms part of the eastern coast of the United States •feeds into the Atlantic Ocean •estuary of the Susquehanna River •separates the eastern and western portions of Maryland and Virginia

Sounds:
    Puget: •forms part of the northwestern coast of the United States •feeds into the Pacific Ocean •located in Washington

Straits:
    Bering : •forms part of the northwestern coast of Alaska •separates Alaska and Russia •connects the Chukchi and Bering Seas
    Strait of Juan de Fuca: •forms part of the northwestern coast of the United States •separates the United States from Canada •connects Puget Sound and the Pacific Ocean
    Straits of Florida: •form part of the southeastern coast of the United States •separate the Florida Peninsula and Cuba •connect the Gulf of Mexico and Atlantic Ocean

Lakes:
    Great Salt: •located in the state of Utah •between the Rocky Mountains and the Great Basin •largest lake in Utah
    Lake Erie: •located in the northeastern United States between Lake Huron and Lake Ontario •on the United States–Canada border •fourth largest of the Great Lakes in surface area
    Lake Huron: •located on the United States–Canada border •third largest of the Great Lakes in surface area •located between Lake Erie and Lake Michigan
    Lake Michigan: •located in the north-central United States between Lake Huron and Lake Superior •second largest of the Great Lakes in surface area •only one of the Great Lakes entirely in the United States
    Lake Okeechobee: •located in the southeastern United States •largest lake in Florida •on the edge of the Everglades
    Lake Ontario: •located in the northeastern United States •on the United States–Canada border •easternmost and smallest of the Great Lakes in surface area •empties into the St. Lawrence River
    Lake Pontchartrain: •located in the southern United States •on the Gulf Coastal

Plain •largest lake in Louisiana
Lake Superior: •located on the United States-Canada border •westernmost and largest of the Great Lakes •largest lake in North America

Rivers:
Arkansas: •tributary of the Mississippi River •has its source in the Rocky Mountains
Colorado: •has its source in the Rocky Mountains •empties into the Gulf of California •extends into Mexico •forms the Grand Canyon
Columbia: •forms much of the Washington-Oregon border •has its mouth in the Pacific Ocean •extends into Canada •a major source of hydroelectric power for the Pacific Northwest
Mississippi: •has its mouth in the Gulf of Mexico •has its source in Lake Itasca in Minnesota •including its tributary the Missouri River, the Mississippi forms the longest river in North America
Missouri: •tributary of the Mississippi River •has its source in the Rocky Mountains •along with the Mississippi River, it forms the longest river in North America
Ohio: •tributary of the Mississippi River •has its source in the Appalachian Mountains
Rio Grande: •has its mouth in the Gulf of Mexico •has its source in the Rocky Mountains •forms much of the United States-Mexico border
Saint Lawrence: •has its mouth in the Gulf of Saint Lawrence •forms part of the United States–Canada border
Snake: •tributary of the Columbia River •has its source in the Rocky Mountains •forms part of the Idaho-Oregon border •forms Hells Canyon, deepest gorge in the United States

Deltas:
Mississippi River Delta: •mouth of the Mississippi River •feeds into the Gulf of Mexico •located in Louisiana

Waterfalls:
Niagara: •located in northeastern United States •part of the Niagara River •on the United States–Canada border (between New York State and Ontario)
Yosemite: •located in western United States •in the Sierra Nevada

Dams:
Glen Canyon: •located on the Colorado River •near the Arizona-Utah border •creates Lake Powell •a hydroelectric dam

Grand Coulee: •located in Washington State •on the Columbia River •a hydroelectric dam
Hoover: •located on the Colorado River •on the Arizona-Nevada border •creates Lake Mead •a hydroelectric dam

Peninsulas:
Alaska: •bordered by the Bering Sea and Pacific Ocean
Florida: •bordered by the Atlantic Ocean, Gulf of Mexico, and Straits of Florida
Olympic: •bordered by the Pacfic Ocean, Strait of Juan de Fuca, and Puget Sound
Seward: •borders the Chukchi Sea, Bering Sea, and Bering Strait

Capes:
Cape Cod: •bordered by the Gulf of Maine and the Atlantic Ocean

Cape Flattery: •bordered by the Pacific Ocean and the Strait of Juan de Fuca •located at the northwestern tip of the Olympic Peninsula •in Washington

Cape Hatteras: •chain of barrier islands in the Atlantic Ocean •off the coast of North Carolina

Islands:
Aleutian: •chain of volcanic islands •extend west from Alaska across the 180-degree meridian •bordered by the Bering Sea and Pacific Ocean

Hawaiian: •island group surrounded by the Pacific Ocean •Hawaii, Kauai, Oahu, Molokai, and Maui are the main islands in the group •volcanic in origin •make up the only island U.S. state

Long: •located off the northeastern coast of the United States •bordered by the Atlantic Ocean and Long Island Sound •part of New York

POLITICAL:
Independence: July 4, 1776 (from Great Britain)
Bordering Countries: Canada, Mexico (2)
Administrative Divisions: Alabama, Alaska, Arizona, Arkansas, California, Colorado, Connecticut, Delaware, Florida, Georgia, Hawaii, Idaho, Illinois, Indiana, Iowa, Kansas, Kentucky, Louisiana, Maine, Maryland, Massachusetts, Michigan, Minnesota, Mississippi, Missouri, Montana, Nebraska, Nevada, New Hampshire, New Jersey, New Mexico, New York, North Carolina, North Dakota, Ohio, Oklahoma, Oregon, Pennsylvania, Rhode Island, South Carolina, South Dakota, Tennessee, Texas, Utah, Vermont, Virginia, Washington, West Virginia, Wisconsin, Wyoming (50 states) •District of Columbia (1 federal district)
External Territories: American Samoa, Baker Island, Guam, Howland Island, Jarvis Island, Johnston Atoll, Kingman Reef, Midway Islands, Navassa Island, Northern Mariana Islands, Palmyra Atoll, Puerto Rico, U.S. Virgin Islands, Wake Island (14)
Ethnic/Racial Groups: white, Hispanic, black, Asian, Amerindian
Religions: Christianity (Protestant, Roman Catholic), Judaism
Languages: English, Spanish
Currency: U.S. dollar
Current President: Barack Obama
Cities (capital, largest, or with at least a million people):
New York: •located in the northeastern United States •city in New York State •most populated city in the United States (8,175,000 people) •chief port on the Atlantic Ocean

Los Angeles: •located in the southwestern United States •largest city in California •on the edge of the Coast Ranges •a major port on the Pacific Ocean

Chicago: •located in the midwestern United States •largest city in Illinois •chief port on Lake Michigan

Houston: •located in the southern United States •city in Texas •a major port on the Gulf of Mexico

Philadelphia: •located in eastern United States •city in Pennsylvania •a major river port on the East Coast

Phoenix: •located in the southwestern United States •city in Arizona •on the edge of the Sonoran Desert
San Diego: •located in the southwestern United States •city in California •a major ocean port and military base on the Pacific Ocean
San Antonio: •located in the south-central United States •city in Texas
Dallas: •located in the south-central United States •city in Texas •on the edge of the Great Plains
Washington, D.C.: •located in the eastern United States •city between Maryland and Virginia •capital of the United States (4,421,000 people)

ENVIRONMENTAL/ECONOMIC:
Climate: mostly temperate; tropical in Hawaii and Florida, arctic in Alaska, semi-arid in the Great Plains, arid in the Great Basin
Natural Resources: coal, copper, lead, molybdenum, phosphates
Agricultural Products: wheat, corn, other grains, fruits, beef
Major Exports: capital goods, automobiles, industrial supplies and raw materials, consumer goods, agricultural products
Natural Hazards: hurricanes, tornadoes, mudslides, forest fires, floods, volcanoes, earthquakes, tsunamis, permafrost (in Alaska)

URUGUAY
Country Name: Oriental Republic of Uruguay
Continent: South America
Area: 68,037 sq mi / 176,215 sq km
Population: 3,369,000
Population Density: 50 people per sq mi / 19 people per sq km
Capital: Montevideo

PHYSICAL:
Highest Point: Cerro Catedral 1,686 ft / 514 m
Lowest Point: Atlantic Ocean 0 ft / 0 m (sea level)
Oceans:
　Atlantic: •forms the eastern coast of Uruguay
Rivers:
　Negro: •tributary of the Uruguay River •extends into Brazil
　Uruguay: •forms all of the Uruguay-Argentina border •feeds into the Rio de la Plata estuary •extends into Brazil
Lakes:
　Lago Rincon del Bonete: •located in central Uruguay •fed by the Negro River
　Lagoa Mirim: •located in eastern Uruguay •extends into Brazil
Deltas:
　Rio de la Plata estuary: •mouth of the Uruguay River •ends in the Atlantic

Ocean •forms most of the southern coast of Uruguay •borders Uruguay and Argentina

POLITICAL:
Independence: August 25, 1825 (from Brazil)
Former Names: Banda Oriental, Cisplatine Province
Bordering Countries: Brazil, Argentina (2)
Administrative Divisions: Artigas, Canelones, Cerro Largo, Colonia, Durazno, Flores, Florida, Lavalleja, Maldonado, Montevideo, Paysandu, Rio Negro, Rivera, Rocha, Salto, San Jose, Soriano, Tacuarembo, Treinta y Tres (19 departments)
Ethnic/Racial Groups: white, mestizo, black
Religion: Christianity (primarily Roman Catholic)
Languages: Spanish, Portunol, Brazilero
Currency: Uruguayan peso
Current President: Jose "Pepe" Mujica Cordano
Cities (capital, largest, or with at least a million people):
   Montevideo: •located in southern Uruguay •city in the Montevideo department •capital of Uruguay •most populated city in Uruguay (1,633,000 people) •chief port on the Rio de la Plata estuary

ENVIRONMENTAL/ECONOMIC:
Climate: warm temperate; temperature almost never falls below freezing
Natural Resources: arable land, hydropower, minor minerals, fisheries
Agricultural Products: rice, wheat, corn, barley, livestock

Major Exports: meat, rice, leather products, wool, vehicles

Natural Hazards: pamperos (violent winds), droughts, floods

UZBEKISTAN
Country Name: Republic of Uzbekistan
Continent: Asia
Area: 172,742 sq mi / 447,400 sq km
Population: 28,463,000
Population Density: 165 people per sq mi / 64 people per sq km
Capital: Tashkent

PHYSICAL:
Highest Point: Adelunga Toghi 14,111 ft / 4,301 m
Lowest Point: Sarygamysh Koli 39 ft / 12 m below sea level
Peaks (minimum elevation of 8,000 ft / 2,400 m):
  Adelunga Toghi: •located in eastern Uzbekistan •on the Uzbekistan-Kyrgyzstan border •highest point in Uzbekistan
Plateau Regions:
  Ustyurt: •located in northwestern Uzbekistan •extends into Kazakhstan
Deserts:
  Qizilqum: •covers much of central Uzbekistan •extends into Kazakhstan
Lakes:
  Aral Sea: •located in northwestern Uzbekistan •extends into Kazakhstan •east of the Ustyurt Plateau and west of the Qizilqum desert

Rivers:

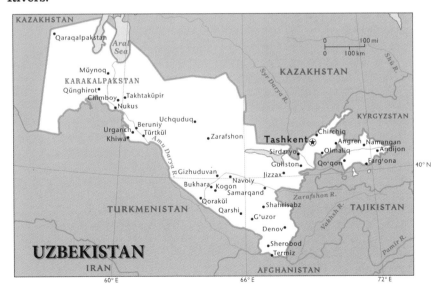

Amu Darya: •has its mouth in the Aral Sea •forms all of the Uzbekistan-Afghanistan border •forms part of the Uzbekistan-Turkmenistan border •extends into Turkmenistan and Afghanistan

Syr Darya: •extends into Tajikistan and Kazakhstan

POLITICAL:
Independence: September 1, 1991 (from the Soviet Union)
Former Name: Uzbek Soviet Socialist Republic
Bordering Countries: Kazakhstan, Turkmenistan, Tajikistan, Kyrgyzstan, Afghanistan (5)
Administrative Divisions: Andijon, Buxoro, Fargona, Jizzax, Namangan, Navoiy, Qashqadaryo, Samarqand, Sirdaryo, Surxondaryo, Toshkent, Xorazm(12 provinces) •Qoraqalpogiston (1 autonomous republic) •Tashkent (1 city)
Ethnic/Racial Groups: Uzbek, Russian, Tajik, Kazakh, Karakalpak, Tatar
Religions: Islam, Christianity (primarily Eastern Orthodox)
Languages: Uzbek, Russian, Tajik
Currency: Uzbekistani sum
Current President: Islom Karimov
Cities (capital, largest, or with at least a million people):
Tashkent: •located in eastern Uzbekistan •city in the Tashkent city division •capital of Uzbekistan •most populated city in Uzbekistan (2,201,000 people)

ENVIRONMENTAL/ECONOMIC:
Climate: mostly midlatitude desert; hot summers/mild winters; semi-arid grassland in the east
Natural Resources: natural gas, petroleum, coal, gold, uranium
Agricultural Products: cotton, vegetables, fruits, grain, livestock
Major Exports: cotton, gold, energy products, mineral fertilizers, ferrous metals
Natural Hazards: droughts, sandstorms

COUNTRIES

VANUATU

Country Name: Republic of Vanuatu
Continent: Australia/Oceania
Area: 4,707 sq mi / 12,190 sq km
Population: 252,000
Population Density: 54 people per sq mi / 21 people per sq km
Capital: Port-Vila

PHYSICAL:

Highest Point: Tabwemasana 6,158 ft / 1,877 m
Lowest Point: Pacific Ocean 0 ft / 0 m (sea level)
Oceans:
  Pacific: •forms the eastern coast of Vanuatu
Seas:
  Coral: •forms the western coast of Vanuatu •feeds into the Pacific Ocean
Islands:
  Espiritu Santo: •largest island in Vanuatu •bordered by the Pacific Ocean and Coral Sea

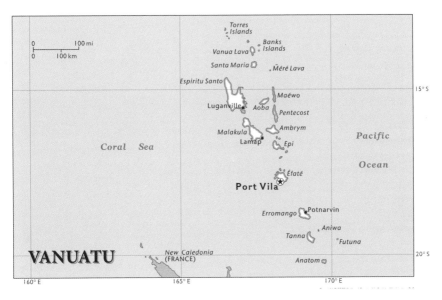

POLITICAL:
Independence: July 30, 1980 (from France and the United Kingdom)
Former Name: New Hebrides
Administrative Divisions: Malampa, Penama, Sanma, Shefa, Tafea, Torba (6 provinces)
Ethnic/Racial Groups: predominantly Melanesian
Religions: Christianity (Protestant, Roman Catholic), indigenous beliefs
Languages: English, French, more than 100 local languages
Currency: vatu
Current President: Iolu Johnson Abbil
Cities (capital, largest, or with at least a million people):
  Port-Vila: •located on Efate Island in central Vanuatu •city in the Shefa province •capital of Vanuatu •most populated city in Vanuatu (44,000 people) •chief port on the Pacific Ocean and the Coral Sea

ENVIRONMENTAL/ECONOMIC:
Climate: tropical, moderated by southeast trade winds
Natural Resources: manganese, hardwood forests, fish
Agricultural Products: copra, coconuts, cocoa, coffee
Major Exports: copra, beef, cacao, timber, kava
Natural Hazards: typhoons, volcanoes, earthquakes, tsunamis

VATICAN CITY
Country Name: State of the Vatican City (The Holy See)
Continent: Europe
Area: 0.2 sq mi / 0.4 sq km (smallest country in the world)
Population: 798 (smallest in the world)
Population Density: 3,990 people per sq mi / 1,995 people per sq km

PHYSICAL:
Highest Point: unnamed location 246 ft / 75 m
Lowest Point: unnamed location 62 ft / 19 m

POLITICAL:
Independence: February 11, 1929 (from Italy)
Bordering Country: Italy (1)
Ethnic/RacialGroups: predominantly Italian and Swiss
Religion: Christianity (predominantly Roman Catholic)
Languages: Italian, Latin, French
Currency: euro
Current Pope: Benedict XVI

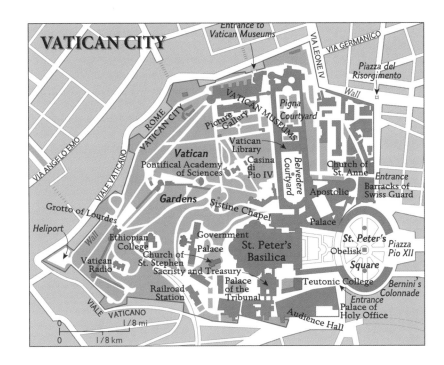

ENVIRONMENTAL/ECONOMIC:
Climate: temperate; mild and rainy winters/hot and dry summers

VENEZUELA
Country Name: Bolivarian Republic of Venezuela
Continent: South America
Area: 352,144 sq mi / 912,050 sq km
Population: 29,278,000
Population Density: 83 people per sq mi / 32 people per sq km
Capital: Caracas

PHYSICAL:
Highest Point: Pico Bolivar 16,427 ft / 5,007 m
Lowest Point: Atlantic Ocean 0 ft / 0 m (sea level)
Mountain Ranges:
    Cordillera de Merida: •located in northwestern Venezuela
    Guiana Highlands: •cover much of southern and central Venezuela •extend into Brazil and Guyana

Peaks (minimum elevation of 8,000 ft / 2,400 m):
  Pico Bolivar: •located in the Cordillera de Merida •highest point in Venezuela
Grasslands/Prairies:
  Llanos: •cover much of western Venezuela •extend into Colombia
Rain forests:
  Amazon: •covers part of southern Venezuela •extends into Colombia and Brazil
Oceans:
  Atlantic: •forms most of the northeastern coast of Venezuela
Seas:
  Caribbean: •forms most of the northern coast of Venezuela •feeds into the Atlantic Ocean
Gulfs:
  Gulf of Paria: •forms part of the northeastern coast of Venezuela •feeds into the Caribbean Sea and the Atlantic Ocean
  Gulf of Venezuela: •forms part of the northwestern coast of Venezuela •feeds into the Caribbean Sea •southern part is called Lake Maracaibo
Lakes:
  Lake Maracaibo: •located in northwestern Venezuela •southernmost part of the Gulf of Venezuela
Rivers:
  Meta: •tributary of the Orinoco River •forms part of the Venezuela-Colombia border •extends into Colombia
  Orinoco: •has its mouth in the Atlantic Ocean •has its source in the Guiana Highlands •forms part of the Venezuela-Colombia border

Deltas:
  Orinoco: •mouth of the Orinoco River •feeds into the Atlantic Ocean
Waterfalls:
  Angel: •located in eastern Venezuela • located in the Guiana Highlands •highest waterfall in the world

POLITICAL:
Independence: July 5, 1811 (from Spain)
Bordering Countries: Brazil, Colombia, Guyana (3)
Administrative Divisions: Amazonas, Anzoategui, Apure, Aragua, Barinas, Bolivar, Carabobo, Cojedes, Delta Amacuro, Falcon, Guarico, Lara, Merida, Miranda, Monagas, Nueva Esparta, Portuguesa, Sucre, Tachira, Trujillo, Vargas, Yaracuy, Zulia (23 states) •Distrito Federal (1 federal district) •Dependencias Federales (1 federal dependency)
Ethnic/Racial Groups: Spanish, Italian, Portuguese, Arab, German, black, indigenous
Religion: Christianity (primarily Roman Catholic)
Languages: Spanish, indigenous dialects
Currency: bolivar
Current President: Hugo Chavez Frias
Cities (capital, largest, or with at least a million people):
  Caracas: •located in northern Venezuela •city in the Distrito Federal and the Miranda state •capital of Venezuela •most populated city in Venezuela (3,051,000 people)
  Maracaibo: •located in northwestern Venezuela •city in the Zulia state •chief port on Lake Maracaibo and the Gulf of Venezuela
  Valencia: •located in northern Venezuela •city in the Carabobo state

ENVIRONMENTAL/ECONOMIC:
Climate: tropical hot and humid; more moderate in the highlands
Natural Resources: petroleum, natural gas, iron ore, gold, bauxite
Agricultural Products: corn, sorghum, sugarcane, rice, beef
Major Exports: petroleum, bauxite, aluminum, steel, chemicals
Natural Hazards: floods, rockslides, mudslides, droughts

VIETNAM
Country Name: Socialist Republic of Vietnam
Continent: Asia
Area: 127,844 sq mi / 331,114 sq km
Population: 87,850,000
Population Density: 687 people per sq mi / 265 people per sq km
Capital: Hanoi

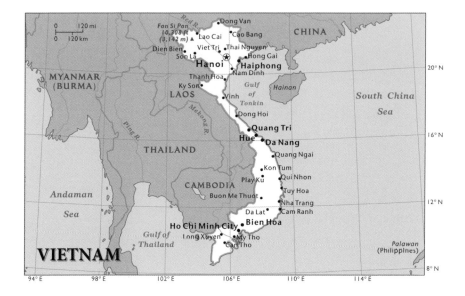

PHYSICAL:
Highest Point: Fan Si Pan 10,308 ft / 3,142 m
Lowest Point: South China Sea 0 ft / 0 m (sea level)
Mountain Ranges:
    Annam Cordillera: •covers much of Vietnam
Peaks (minimum elevation of 8,000 ft / 2,400 m):
    Fan Si Pan: •located in northwestern Vietnam •highest point in Vietnam
Seas:
    South China: •forms much of the southern and eastern coasts of Vietnam •feeds into the Philippine and Java Seas
Gulfs:
    Gulf of Thailand: •forms part of the southwestern coast of Vietnam •feeds into the South China Sea
    Gulf of Tonkin: •forms the northeastern coast of Vietnam •feeds into the South China Sea
Rivers:
    Mekong: •has its mouth in the South China Sea •extends into Cambodia, Laos, and China
    Red: •has its mouth in the Gulf of Tonkin •forms part of the Vietnam-China border •extends into China
Deltas:
    Mekong River Delta: •mouth of the Mekong River •feeds into the South China Sea

POLITICAL:
Independence: September 2, 1945 (from France)
Former Names: North Vietnam, South Vietnam
Bordering Countries: Laos, China, Cambodia (3)
Administrative Divisions: An Giang, Bac Giang, Bac Kan, Bac Lieu, Bac Ninh, Ba Ria-Vung Tau, Ben Tre, Binh Dinh, Binh Duong, Binh Phuoc, Binh Thuan, Ca Mau, Cao Bang, Dac Lak, Dac Nong, Dien Bien, Dong Nai, Dong Thap, Gia Lai, Ha Giang, Ha Nam, Ha Tinh, Hai Duong, Hau Giang, Hoa Binh, Hung Yen, Khanh Hoa, Kien Giang, Kon Tum, Lai Chau, Lam Dong, Lang Son, Lao Cai, Long An, Nam Dinh, Nghe An, Ninh Binh, Ninh Thuan, Phu Tho, Phu Yen, Quang Binh, Quang Nam, Quang Ngai, Quang Ninh, Quang Tri, Soc Trang, Son La, Tay Ninh, Thai Binh, Thai Nguyen, Thanh Hoa, Thua Thien-Hue, Tien Giang, Tra Vinh, Tuyen Quang, Vinh Long, Vinh Phuc, Yen Bai (58 provinces) •Can Tho, Da Nang, Ha Noi, Hai Phong, Ho Chi Minh City (5 municipalities)
Ethnic/Racial Groups: Vietnamese, Chinese, Hmong, Thai, Khmer, Cham
Religions: Buddhism, Hoa Hao, Cao Dai, Christianity, indigenous beliefs, Islam
Languages: Vietnamese, English, French, Chinese, Khmer, local languages
Currency: dong
Current President: Trong Tan Sang
Cities (capital, largest, or with at least a million people):
    Ho Chi Minh City: •located in southern Vietnam •city in the Ho Chi Minh municipality •most populated city in Vietnam (6,167,000 people)
    Hanoi: • located in northern Vietnam •city in the Hanoi municipality •capital of Vietnam (2,668,000 people) •chief port on the Red River

ENVIRONMENTAL/ECONOMIC:
Climate: tropical in the south; monsoonal in the north; rainy/dry seasons
Natural Resources: phosphates, coal, manganese, bauxite, chromate
Agricultural Products: paddy rice, corn, potatoes, rubber, poultry
Major Exports: crude oil, marine products, rice, coffee, rubber
Natural Hazards: typhoons, floods

YEMEN
Country Name: Republic of Yemen
Continent: Asia
Area: 207,286 sq mi / 536,869 sq km
Population: 23,833,000
Population Density: 115 people per sq mi / 44 people per sq km
Capital: Sanaa

PHYSICAL:
Highest Point: Mount Nabi Shuayb 12,336 ft / 3,760 m
Lowest Point: Arabian Sea 0 ft / 0 m (sea level)
Mountain Ranges:
 Hadramawt: •covers much of southern Yemen
 Jabal al Hijaz: •covers much of western Yemen •extends into Saudi Arabia
Peaks (minimum elevation of 8,000 ft / 2,400 m):
 Mount Nabi Shuayb: •located in the Jabal al Hijaz •highest point in Yemen
Deserts:
 Rub al Khali: •covers much of northern Yemen •extends into Saudi Arabia and Oman

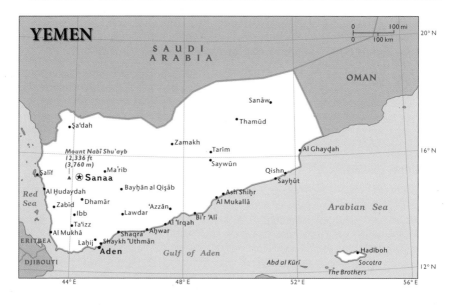

Seas:
  Arabian: •forms part of the eastern coast of Yemen •feeds into the Indian Ocean
  Red: •forms much of the western coast of Yemen •feeds into the Gulf of Aden
Gulfs:
  Gulf of Aden: •forms most of the southern coast of Yemen •feeds into the Arabian Sea
Straits:
  Bab al Mandab: •forms part of the southwestern coast of Yemen •separates Yemen from Eritrea and Djibouti •connects the Red Sea and Gulf of Aden
Islands:
  Socotra: •island at the eastern end of the Gulf of Aden •bordered by the Arabian Sea and Gulf of Aden

POLITICAL:

Independence: May 22, 1990 (unification of North and South Yemen)
Former Names: North Yemen, South Yemen
Bordering Countries: Saudi Arabia, Oman (2)
Administrative Divisions: Abyan, Adan, Ad Dali, Al Bayda, Al Hudaydah, Al Jawf, Al Mahrah, Al Mahwit, Amanat al Asimah, Amran, Dhamar, Hadramawt, Hajjah, Ibb, Lahij, Marib, Raymah, Sadah, Sana, Shabwah, Taizz (21 governorates)
Ethnic/Racial Groups: predominantly Arab
Religion: Islam (Sunni, Shiite)
Language: Arabic
Currency: Yemeni rial
Current President: Ali Abdallah Salih
Cities (capital, largest, or with at least a million people):
  Sanaa: •located in western Yemen •city in the Sana governorate •capital of Yemen •most populated city in Yemen (2,229,000 people) •located in the Jabal al Hijaz

ENVIRONMENTAL/ECONOMIC:

Climate: mostly desert; hot and humid along the coast; temperate in the western mountains; hot and dry desert in the east
Natural Resources: petroleum, fish, rock salt, marble, coal
Agricultural Products: grain, fruits, vegetables, pulses, dairy products
Major Exports: crude oil, coffee, dried and salted fish
Natural Hazards: sandstorms, dust storms

ZAMBIA
Country Name: Republic of the Zambia
Continent: Africa
Area: 290,586 sq mi / 752,614 sq km
Population: 13,475,000
Population Density: 46 people per sq mi / 18 people per sq km
Capital: Lusaka

PHYSICAL:
Highest Point: location in the Mafinga Hills 7,549 ft / 2,301 m
Lowest Point: Zambezi River 1,079 ft / 329 m
Mountain Ranges:
  Muchinga: •located in eastern Zambia
Lakes:
  Lake Bangweulu: • located in northern Zambia
  Lake Kariba: •located in southern Zambia •formed by the Kariba Dam •fed by the Zambezi River •on the Zambia-Zimbabwe border
  Lake Mweru: •located in northern Zambia •on the Zambia-Democratic Republic of the Congo border
  Lake Tanganyika: •located in northern Zambia •extends into the Democratic Republic of the Congo and Tanzania
Rivers:
  Luangwa: •tributary of the Zambezi River •has its source in the Muchinga Mountains •forms part of the Zambia-Mozambique border
  Zambezi: •has its source in northwestern Zambia •forms all of Zambia's border with Zimbabwe and Botswana •forms much of the Zambia-Namibia border •extends into Angola and Mozambique
Waterfalls:
  Victoria: •located in southern Zambia •on the Zambezi River •on the Zambia-Zimbabwe border
Dams:
  Kariba: •located in southeastern Zambia •forms Lake Kariba •on the Zambezi River •on the Zambia-Zimbabwe border

POLITICAL:
Independence: October 24, 1964 (from the United Kingdom)
Former Name: Northern Rhodesia

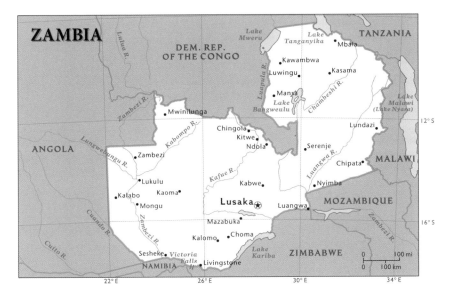

Bordering Countries: Democratic Republic of the Congo, Angola, Malawi, Zimbabwe, Mozambique, Tanzania, Namibia, Botswana (8)
Administrative Divisions: Central, Copperbelt, Eastern, Luapula, Lusaka, Northern, North-Western, Southern, Western (9 provinces)
Ethnic/Racial Groups: more than 70 indigenous groups, including Bamba, Tonga, Ngoni, Lozi
Religions: Christianity, Islam, Hinduism
Languages: English, indigenous languages
Currency: Zambian kwacha
Current President: Michael Chilufya Sata
Cities (capital, largest, or with at least a million people):
　Lusaka: ●located in south-central Zambia ●city in the Lusaka province ●capital of Zambia ●most populated city in Zambia (1,413,000 people)

ENVIRONMENTAL/ECONOMIC:
Climate: tropical, modified by altitude; rainy/dry seasons
Natural Resources: copper, cobalt, zinc, lead, coal
Agricultural Products: corn, sorghum, rice, peanuts, cattle
Major Exports: copper, cobalt, electricity, tobacco, flowers
Natural Hazards: droughts, tropical storms

ZIMBABWE

Country Name: Republic of Zimbabwe
Continent: Africa
Area: 150,872 sq mi / 390,757 sq km
Population: 12,084,000
Population Density: 80 people per sq mi / 31 people per sq km
Capital: Harare

PHYSICAL:

Highest Point: Inyangani 8,504 ft / 2,592 m
Lowest Point: Runde/Save River junction 531 ft / 162 m
Peaks (minimum elevation of 8,000 ft / 2,400 m):
  Inyangani: ●located in eastern Zimbabwe ●highest point in Zimbabwe
Lakes:
  Lake Kariba: ●located in northwestern Zimbabwe ●formed by the Kariba Dam ●on the Zambezi River ●on the Zimbabwe-Zambia border
Rivers:
  Limpopo: ●forms all of the Zimbabwe-South Africa border ●extends into Mozambique
  Zambezi: ●forms all of the Zimbabwe-Zambia border ●extends into Zambia and Mozambique
Waterfalls:
  Victoria: ●located in western Zimbabwe ●on the Zambezi River ●on the Zimbabwe-Zambia border
Dams:
  Kariba: ● located in northwestern Zimbabwe ●forms Lake Kariba ●on the Zambezi River ●on the Zimbabwe-Zambia border

POLITICAL:

Independence: April 18, 1980 (from the United Kingdom)
Former Names: Rhodesia, Southern Rhodesia
Bordering Countries: Mozambique, Botswana, Zambia, South Africa (4)
Administrative Divisions: Manicaland, Mashonaland Central, Mashonaland East, Mashonaland West, Masvingo, Matabeleland North, Matabeleland South, Midlands (8 provinces) ●Bulawayo, Harare (2 cities)
Ethnic/Racial Groups: Shona, Ndebele
Religions: syncretic (part Christian and part indigenous beliefs), Christianity, indigenous beliefs
Languages: English, Shona, Sindebele
Currency: Zimbabwean dollar
Current President: Robert Gabriel Mugabe
Cities (capital, largest, or with at least a million people):
  Harare: ●located in northern Zimbabwe ●city in the Harare city division ●capital of Zimbabwe ●most populated city in Zimbabwe (1,606,000 people)
  Bulawayo: ●located in southwestern Zimbabwe ●city in the Bulawayo city division

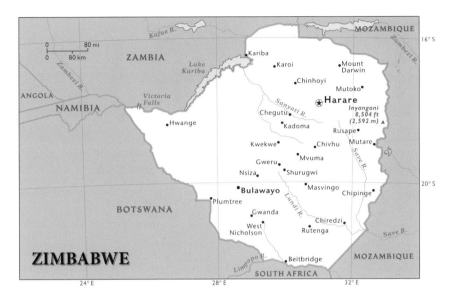

ENVIRONMENTAL/ECONOMIC:
Climate: tropical, moderated by altitude; rainy/dry seasons
Natural Resources: coal, chromium ore, asbestos, gold, nickel
Agricultural Products: corn, cotton, tobacco, wheat, cattle
Major Exports: tobacco, gold, ferroalloys, textiles, clothing
Natural Hazards: droughts, floods, severe storms

Geographic Extremes

Largest in area: Russia
North America: Canada
South America: Brazil
Europe: Russia (Ukraine is the largest country entirely in Europe.)
Asia: Russia (China is the largest country entirely in Asia.)
Africa: Algeria
Australia/Oceania: Australia

Smallest in area: Vatican City
North America: St. Kitts and Nevis
South America: Suriname
Europe: Vatican City
Asia: Maldives
Africa: Seychelles
Australia/Oceania: Nauru

Largest in population: China
North America: United States
South America: Brazil
Europe: Russia (Germany is the most populated country entirely in Europe.)
Asia: China
Africa: Nigeria
Australia/Oceania: Australia

Smallest in population: Vatican City
North America: St. Kitts and Nevis
South America: Suriname
Europe: Vatican City
Asia: Maldives
Africa: Seychelles
Australia/Oceania: Tuvalu

Densest in population: Monaco
North America: Barbados
South America: Ecuador
Europe: Monaco
Asia: Singapore
Africa: Mauritius
Australia/Oceania: Nauru

Least dense in population: Mongolia
North America: Canada
South America: Suriname
Europe: Iceland
Asia: Mongolia
Africa: Namibia
Australia/Oceania: Australia

Highest point: Mount Everest (Nepal/China)
North America: Mount McKinley (U.S.)
South America: Aconcagua (Argentina)
Europe: Mount Elbrus (Russia)
Asia: Mount Everest (Nepal/China)
Africa: Kilimanjaro (Tanzania)
Australia/Oceania: Mt. Wilhelm (Papua New Guinea)

Lowest point: Dead Sea (Israel/Jordan)
North America: Death Valley (U.S.)
South America: Laguna del Carbon (Argentina)
Europe: Caspian Sea (Russia)
Asia: Dead Sea (Israel/Jordan)
Africa: Lake Assal (Djibouti)
Australia/Oceania: Lake Eyre (Australia)

Largest grassland: Eurasian steppe (Europe/Asia)

Largest plateau: Tibetan (China)

Largest rain forest: Amazon (South America)

Largest swamp: Pantanal (South America)

Largest desert: Sahara (Africa)

Largest lake: Caspian Sea (Russia/ Azerbaijan/Iran/Turkmenistan/ Kazakhstan)
North America: Lake Superior (U.S./Canada)
South America: Lake Titicaca° (Peru/Bolivia)
Europe: Caspian Sea (Russia); Lake Ladoga (Russia) is the largest lake entirely in Europe.
Asia: Caspian Sea (Azerbaijan/Iran/ Turkmenistan/Kazakhstan); Lake Baikal (Russia) is the largest lake entirely in Asia.
Africa: Lake Victoria (Kenya/Tanzania/ Uganda)
Australia/Oceania: Lake Eyre (Australia)

Longest river: Nile (Africa)
North America: Mississippi-Missouri
South America: Amazon
Europe: Volga
Asia: Chang Jiang (Yangtze)
Africa: Nile
Australia/Oceania: Murray-Darling

Largest river discharge: Amazon (South America)

Tallest waterfall: Angel Falls (Venezuela)

Largest dam: Three Gorges Dam (China)

Largest island: Greenland

°Lake Maracaibo, although called a lake, is actually an extension of the Gulf of Venezuela.

GEOGRAPHIC EXTREMES | PAGE 379

Glossary

Amerindian: a term for the native peoples of the Americas

animism: the belief that natural objects, such as trees, mountains, and rivers, are inhabited by spirits

Austronesian: people of mixed Australian and Polynesian, Melanesian, Micronesian, Malaysian, or Formosan ancestry

capital goods: tools, machinery, and other equipment or items used to produce goods and services

Carib: a descendant of Amerindian people who formerly lived in certain islands of the West Indies and in northern South America; also the language of these people

CIS: an acronym for the Commonwealth of Independent States, an alliance established in 1991 and made up of republics of the former Soviet Union (Web sites: www.cis.minsk.by / www.cisstat.com/eng/cis.htm)

continental climate: a midlatitude climate occurring on large landmasses in the Northern Hemisphere; characterized by great variations in temperature seasonally and between day and night

Coptic: the native Christian religion of Egypt and Ethiopia

creole: a person of European heritage, especially French, born in the West Indies, Central America, tropical South America, or the U.S. Gulf States

erg: a huge desert area characterized by deep sand dunes

estuary: the broadened seaward end of a river that usually has a mixture of fresh and salt water that rises and falls with the tides

ethnic group: a group of people with a common ancestry and cultural tradition; examples include Irish, Hispanic, and Russian

Euronesian: people of mixed European and Polynesian, Micronesian, or Melanesian ancestry

ferroalloy: an alloy of iron used in the manufacture of steel, such as chromium and manganese

foehn: a warm, dry wind blowing into mountain valleys, especially in the European Alps

geothermal energy: heat energy generated from Earth's interior

harmattan: a hot, dusty wind that blows from the Sahara to the Atlantic coast of Africa, primarily from November to March

manufactured goods: raw materials that have been made into marketable goods by the use of industrial equipment

Melanesian: people indigenous to Melanesia, the part of Oceania that includes the islands of Fiji, New Guinea, Vanuatu, the Solomon Islands, and New Caledonia

mestizo: a person, especially in Latin America, whose ancestry is a mix of Spanish or Portuguese and American Indian

Micronesian: people indigenous to Micronesia, the part of Oceania that includes Guam, Kiribati, the Mariana Islands, Palau, and the Federated States of Micronesia

mistral: a violent wind blowing south to the Mediterranean coast of France, most often in winter and spring, that can last for a hundred days each year

monsoon: a seasonal change in the direction of prevailing winds that causes wet and dry seasons in some areas

mulatto: a person of mixed black and white ancestry

Negrito: a term for any dark-skinned person of short stature living in Oceania or Southeast Asia

nonferrous metals: metals that contain no iron

Oceania: name for the widely scattered islands of Polynesia, Micronesia, and Melanesia; often includes Australia and New Zealand

patois: a form of a language that differs from the accepted standard; a local dialect

permafrost: a permanently frozen layer of soil, gravel, and sand bound together by ice; found below Earth's surface at high latitudes or elevations

pidgin English: a form of English that is a mix of English and one or more local languages

Polynesian: people indigenous to Polynesia, the part of Oceania that includes the Hawaiian Islands, the Society Islands, Samoa, and French Polynesia

precious metals: relatively scarce but valuable metals, including gold, silver, platinum, and iridium

racial group: a group of people defined by its genetic makeup; examples include whites, blacks, and Amerindians

re-exports: foreign goods that are exported by a nation in the same form in which they were originally imported into that nation

Sahel: a semiarid region of short, tropical grassland that lies between the Sahara and the tropical wet region of equatorial Africa and that is prone to frequent droughts

sirocco: a hot, usually springtime wind blowing from the Sahara to the Mediterranean coast of Africa

Soviet Union: an abbreviated name for the former Union of Soviet Socialist Republics (U.S.S.R.) that consisted of 15 republics, the largest of which was the Russian Soviet Federated Socialist Republic (Russia); ceased to exist in 1991

steppe: a Slavic word for the relatively flat, mostly treeless temperate grasslands that stretch across much of central Europe and central Asia

steppe climate: associated with midlatitude grasslands; characterized by hot summers, cold winters, and moderate rainfall

tsunami: powerful and potentially destructive ocean waves triggered by an earthquake or volcanic eruption

Bibliography

National Geographic Atlas of the World, 9th Edition. National Geographic Society, Washington, D.C., 2010.

National Geographic Xpeditions Atlas. National Geographic Society, Washington, D.C., www.nationalgeographic.com/xpeditions/atlas

The World Factbook. Central Intelligence Agency, Washington, D.C. www.cia.gov/library/publications/the-world-factbook/index.html

The World Gazetteer. 21 Jun 2004 www.world-gazetteer.com

Andrew Wojtanik is a 22-year-old senior at Georgetown University, where he is studying international security with a minor in African studies. Eight years after winning the National Geographic Bee, he continues to enjoy looking at maps, charting out territory, and traveling to the destinations he has studied. Having worked for eight months at the U.S. State Department, he hopes to pursue a career in international diplomacy. Andrew also enjoys hiking, backpacking, and cheering for the Buffalo Bills and Georgetown Hoyas. He currently lives in Washington, D.C., but the future remains an exciting mystery.

Your Notes